DeWALT®

BUILDING CONTRACTOR'S LICENSING

EXAM GUIDE

Chris Prince
American Contractors Exam Services

Created exclusively
for DeWALT by:

www.dewalt.com/guides

ISBN-10: 0-9770003-7-0
ISBN-13: 978-0-9770003-7-1

Pal Publications, Inc.
800 Heritage Drive
Suite 810
Pottstown, PA 19464

Titles Available From DeWALT:

DeWALT Trade Reference Series

Blueprint Reading Professional Reference

Construction Professional Reference

Construction Estimating Professional Reference

Construction Safety/OSHA Professional Reference

Datacom Professional Reference

Electrical Professional Reference

Electrical Estimating Professional Reference

Electric Motor Professional Reference

HVAC/R Professional Reference – Master Edition

Lighting & Maintenance Professional Reference

Plumbing Professional Reference

Residential Remodeling & Repair Professional Reference

Security, Sound & Video Professional Reference

Spanish/English Construction Dictionary – Illustrated Edition

Wiring Diagrams Professional Reference

DeWALT Exam and Certification Series

Electrical Licensing Exam Guide

HVAC Technician Certification Exam Guide

Plumbing Licensing Exam Guide

CONTENTS

DEWALT BUILDING CONTRACTOR'S LICENSING EXAM GUIDE

by
Chris Prince
American Contractors Exam Services

INTRODUCTION

Congratulations! If you are reading this book, it is likely that you are pursuing new opportunity. Whether you are preparing to take an exam in order to start your own contracting business, to enhance the qualifications of an existing company, or to qualify the company for which you now work, you are taking deliberate measures to improve an existing situation. So sit back, relax, and rest assured that the time you spend with this book will significantly improve your chances of passing the exam that you are about to take.

TIME FOR TEST MODE!

Preparing for and taking an exam requires a specific mind-set. It requires planning for the upcoming event and making a conscious effort to control anxiety. For most, the thought of taking an exam will cause a feeling of butterflies churning in the stomach. For some, it will lead to more severe symptoms such as headaches and nausea. Don't fear; controlled test anxiety is good. It creates a sense of urgency to succeed. Identify your level of anxiety, and control it.

How do you rate your level of anxiety? For some, perhaps the last test-taking ordeal was a high school final exam twenty years or so ago. The memory of this experience, coupled with the pressure of living up to the expectations of family, a boss, or coworkers can be overwhelming. When children are aware that a parent will soon be taking an exam, they are likely to reciprocate the pressure applied to do well in school. Bosses often assume that the exam will be a breeze for such a competent employee, and many times, a spouse's support is mistaken for unrealistic expectation. Peer pressure is not necessary. Do what you can to minimize it.

Keep the fact that you are taking the exam a secret! Maybe everyone notices your study efforts, but there is no need to announce the test date. Make your plans by scheduling the exam, and if you must explain, say that you have an appointment. The pressures felt from the need to please and live up to the expectations of everyone around you can lead to undue stress and is not necessary. If no one knows your plans, the only pressure you will feel to succeed is your own. This pressure is healthy and probably necessary. Rather than dealing with the pressure of others, imagine the feeling when you announce your accomplishment after the fact. Work toward this goal, and visualize success. Think positive.

Another way to eliminate stress may be to think of your first test date as a practice run. In the worst-case scenario, passing the exam will take two attempts. It is not the end of the world if you do not pass on the first attempt. Most states and municipalities allow you to take the exam as many times as necessary without penalty. Before using this suggestion, make sure that your exam does not limit the number of times it can be taken during a certain amount of time. (We will explain later how to check this.)

Notes

Next, avoid becoming irritated that the exam is required. Forget the fact that you know your trade and feel that you should not be required to pass an exam to prove yourself. Look at the experience as an opportunity. The knowledge gained from preparing for the exam will far outweigh the effort. Without a doubt, you will gain information that will improve your efficiency and trade expertise, and could even save you money.

Remember the old carpenter's advice to measure twice and cut once? Similarly, when it comes to test-taking, it is critical to be precise, careful, and methodical as you prepare for exam day. Apply this philosophy as you take the practice exams in this book—read the questions twice and choose the answers carefully. The exams are typically administered in a multiple-choice format, and it is not uncommon for at least two of the choices to appear to be correct, depending on how the question is read. Remember, there is only one correct answer, and one word in the question can change its entire meaning. Be careful.

After you have switched to test mode, it is time to prepare for the exam.

PREPARING FOR THE EXAM

The first step of preparation is to identify the exam required to accomplish your goal. If you are taking a state-required exam to become a licensed contractor, you can begin by visiting www.becomealicensedcontractor.com. Select your state, choose General Contracting under the listing of test types, and the test requirements will be found. The site will likely lead you to the candidate information bulletin provided by the approved testing company.

The candidate information bulletin is the test-taker's rule book. It provides specific information, including the addresses of testing centers, identification requirements, security procedures, and anything else necessary for taking the exam. In preparing for the exam, it is important to identify the answers to the following questions when using the candidate information bulletin.

1. ***What are the approved references?*** This is, by far, the most important piece of information to know prior to your exam preparation mission. Most general contracting exams are based on either the International Residential Code or the International Building Code and several additional references. After identifying the approved reference materials, eliminate the study exams in this book that are not included in your exam. Do not spend valuable time studying information that is not included on your exam.

Notes

2. ***Are the approved references allowed in the examination center?*** In most cases, books are allowed in the examination room for use throughout the exam. While this would seem to make the exam extremely easy, if you are not familiar with how to use the books, it might as well be a closed-book exam.

3. ***Is tabbing and highlighting of the books allowed?*** Typically, if the approved references are allowed to be taken into the exam room, they are allowed to be highlighted and tabbed prior to test day. Many rules allow you to underline but prohibit you from making notes in the reference materials. The candidate information bulletin will specifically call for permanent tabs. Permanent tabs are those that cannot be easily removed. Post-it tabs are generally not allowed. If any of the rules are broken, your reference materials may be banned from the test site.

4. ***What is the time frame for taking the exam?*** The time frame is important because you will want to simulate the time allowed as you work through the study exams throughout this book. Simply take the allotted time in minutes and divide by the number of questions on the exam to determine the time allotment per question. You can now easily multiply the number of questions in a particular section of this book by the average time allowed per question to set the time frame for working through the practice exams. In the beginning, you will want to allot additional time until you become familiar with the reference materials.

5. ***What is the content outline of the exam?*** It is important to pay attention to the breakdown of the exam in order to allot your study time. If only five out of 80 questions pertain to OSHA, you should spend less time studying this subject and concentrate more on the areas representing the largest portion of the exam.

USING THIS EXAM GUIDE

Before using this guide, make sure that you have highlighters and tabs readily available. Begin with identifying the table of contents, the index, and the glossary of each reference book. Place a tab on each of these sections for easy access. The table of contents divides the book into chapters or subjects and will be used frequently. The index is an alphabetical listing of key words found throughout the book and should be your starting point for finding an answer to a specific question. If the answer is not found using the index, identify the chapter according to the subject of the question by using the table of contents. The glossary is an alphabetized listing of terms that serves as a useful source for answering questions

Notes

quickly. A glossary may or may not be found in the approved references.

Each section of this exam guide is based on a particular subject. As you answer each of the questions from the study exam, highlight the answers. This will help you to become familiar with the book and will strengthen your ability to quickly reference important code items. Remember, the purpose of each practice exam is not to test your trade knowledge; it is to provide an exercise of how to navigate and use the approved reference(s). If you feel that you know the answer to a question, you should verify the answer using the applicable book. Pay close attention to the tables that are referenced as well as the subject matter of each chapter.

When it comes to tabbing your reference materials, be careful not to overdo it. Placing too many tabs will be more of a hindrance than a useful tool. Remember, when using tabs, the more tabs you use, the longer it will take to read through each one. Using the index will likely save you time in your search for specific information. It is highly recommended that you tab the chapters, table of contents, index, and useful tables.

Going into the exam, make sure that you are familiar with the subject matter of each book. It is imperative that you understand the layout and format of each of the references. This is accomplished through the process of answering the questions in this guide.

Prior to exam day, remove the index from approved reference books that are bound with a three-ring binder, staple it together, and place it in the front pocket of the binder. This will allow easy access to the index by being able to place it on the testing table beside the reference book rather than requiring you to flip back and forth between the index and the body of the book.

On average, exams allow about 3 minutes per question. When you begin using the study exams in this book, allow yourself 8 to 10 minutes per question. As you become more familiar with the reference materials, decrease this time allotment to about 5 minutes. By the time you complete the final exams, only allow yourself the time allotted on your exam.

It is important to devise a time management strategy that works for you immediately. By test day, your goal is to have a plan of action of how to work through the exam. As you are preparing for the exam, keep in mind that the clock will be working against you. Decide the maximum amount of time that you will spend on one question before moving on to the next. You do not want to run out of time.

Notes

WHAT TO BRING TO THE EXAM

Prior to test day, make sure you get a good night's rest and arrive armed with the following items:

Candidate information bulletin. Many times, the proctor of the exam is inexperienced. If you are told that tabs are not allowed, you need to be able to defend yourself by referring to the bulletin (the rule book).

Bottle of water. The clock does not stop during the exam. There are no "hold" buttons. If you need a sip of water and have to run to the water fountain or bathroom, the timer will continue to count down. Although you have little choice when it comes to a bathroom break, at least be prepared for the dry mouth syndrome.

Magnifying glass. Many of the documents and diagrams used throughout the exam are difficult to read. Save the frustration and headache-causing eyestrain, and use a magnifying glass.

Two pencils and a pencil sharpener. Arrive with at least two pencils and a pencil sharpener, especially if you are taking the exam using the old-fashioned pencil-and-paper format. Tests are still administered this way in several states and municipalities.

Two calculators. Remember Murphy's law—if it can go wrong it probably will on test day. Have a contingency plan for everything. If you insist on using your favorite calculator, and it happens to be one that you are not sure is allowed, such as a construction master, make sure that you have a backup.

A great attitude. Make every attempt to remain calm, cool, and collected. This is easier to maintain if you have had adequate rest the night before the exam. Do not cram and stay up until midnight. This will work against your tolerance level for aggravation. Remember, frustration will only create tension and make everything more difficult.

Make sure that you also understand the requirements for identification, payment methods, and proper exam registration documentation.

LET THE TEST BEGIN

The moment you sit in the "hot seat" to begin the exam, let the strategy unravel. Have a plan and stick with it. A few recommendations:

1. *Switch to test mode.* Many times, the rules of thumb and assumptions you make in the field will not work in the test world. While your background and experience can be an attribute, do

Notes

not allow it to get in the way on test day. Make no assumptions. If you are not 100% sure of an answer, try to verify it at some point. Be deliberate to focus and concentrate on each question.

2. ***Arrange your work area neatly.*** Stack your books to one side and, if possible, stand them upright so that they are easily accessible. Decide for yourself that at least for this day, you will be the most organized and careful person in the world.

3. ***Download the memory.*** That's right, download the information you are having trouble remembering. Transfer the formulas and anything else you have been repeating since you walked into the test site from your brain to the scratch sheet of paper provided by the proctor.

4. ***Become familiar with the construction drawings and/or diagram booklet before you start the clock.*** If your test is being administered by computer, you are in control of when the countdown begins. Take advantage of the control, but be careful not to push it. The proctor is only a few steps away and may prod you to begin the test if you wait too long.

5. ***Answer the easy questions first.*** Nothing will boost your confidence more than to run through a good portion of the exam answering questions based on information you recall from your studies. In contrast, your confidence level can diminish rapidly if you become distracted by a difficult question.

6. ***"Mark" any questions answered that you doubt.*** If the test is computer based, you have the option to mark questions to review later. If you run out of time, the computer accepts the selected answer and does not penalize you for marking the question. By marking questions, if you have additional time after answering all questions, you can verify the answers you selected.

7. ***Leave the difficult questions unanswered, and come back to these last.*** As you skip questions, make a note on the scratch paper, associating the question number with a specific book. This will allow you to categorize each of the unanswered questions by book, saving valuable time and unnecessary frustration. After reading the last question, you will be allowed to revisit unanswered questions or marked questions. You can choose to go to the first of these in the group or to a specific question number.

8. ***Do not run out of time.*** Pay attention to the clock. Do not leave any question unanswered. Before you run out of time, select a

Notes

choice for each question. Questions left unanswered will be counted against you. If you have to guess on a number of questions, improve your odds by selecting the same choice on each question. For example, select all "a's" or all "c's," but your goal is to manage your time and not have to guess on any of the questions.

PART ONE
Math Concepts

Understanding basic math concepts can be crucial for the portion of the exam, which is based on print reading.

If your candidate information bulletin does not list math or plan reading as subjects to be covered, this chapter may not be as important to you as some of the others. However, it will prove beneficial for shifting your thought process to test mode. Working through the exercises will stimulate your brain. This chapter can be thought of as the "warm up" recommended for most athletes prior to an intense workout.

While many contractors probably dread math as much as many Americans dread exercise, it is as crucial to running a successful business as exercise is to health. What many exam candidates come to realize is that math is not nearly as bad as what they have come to expect. In fact, most people are better at math than they think.

The following chapter covers basic math concepts that are crucial to the contracting business and, more importantly, for now at least, crucial to passing an exam. While it is impossible to cover every type of question that may appear on an exam, a strong knowledge of the proper steps to performing similar math calculations will get you through any exam.

Concepts covered in this chapter include:

- Math rules
- Rounding
- Converting inches to feet
- Square feet to estimate brick, block, and drywall
- Cubic yards to estimate concrete and dirt
- Framing math
- Roofing math

THE DREADED MATH

When it comes to math portions of the exam, there are several important and critical rules. It is in your best interest to convince yourself that for this one day, you are going to take a methodical approach and deliberately follow each of the rules that follow.

MATH RULES

1. ***Always use a calculator.*** This is probably the most critical rule when it comes to performing math operations on the exam. Never perform mental calculations. Once you arrive at an answer, do the calculation again to check your answer. Repeat your calculations until you are certain the answer is correct.

2. ***Use scratch paper to write down each step of the math.*** Professionals who use math on a daily basis, accountants, engineers, and the like, are sometimes the most difficult math students. Why? Because they simply refuse to use scratch paper to write each step of the equation as they work through it. They are also reluctant to use a calculator for what they believe to be simple math. You do not want to fail this exam by one or two points; it hurts worse to make a grade of 69 than it does to make a grade of 30. The point is, don't get in a hurry and miscalculate. Always write your problem out as you solve it. Always perform math operations, at least twice, with a calculator.

3. ***Always multiply "like" numbers.*** In other words, do not mix feet and inches. If you multiply 10 feet by 8 feet 6 inches, the answer is not 86 feet, it is 85 feet. You cannot simply multiply 10 by 8.6—you must first convert the 6 inches to feet (to be reviewed later).

4. ***When converting inches to feet, round to the nearest hundredth.*** Yes, this is an exception to Rule 3. The reason we round to the nearest hundredth when converting inches to feet is because it puts your answer in the correct range of the choices provided and allows you to write it down more easily. You would not want to write 0.3333333333333 when it would be much simpler to write 0.33 (to be reviewed later).

5. ***Never round until you get to the end.*** This simply means that you should never round any of your numbers until you solve the equation. When necessary, round based on the choices provided to a question or to the directions provided in the question. The only ***exception*** to this rule is when converting inches to feet (Rule #4).

Example:

Rounding up to the next whole yard, how many cubic yards of concrete will be needed to pour a 4-inch-deep drive that measures 12 feet × 100 feet?

a. 24
b. 24.44
c. 25
d. 25.5

Solution:

$$CY = \frac{L \times W \times D}{27}$$

$$CY = \frac{100 \times 12 \times .33}{27}$$

$$CY = 24.44$$

Explanation:

The question clearly states "round to the next whole yard." Although the exact answer is one of the choices, 25 (choice c) is the correct answer. Notice that 4 inches was converted to feet allowing the multiplication of "like" numbers. Before solving the equation, 0.33 was rounded. This is acceptable and will put you in the proper range of the correct answer for your exam.

6. ***Draw it out.*** If you have to sketch apples and stick people, do it. Remember, you do not want to fail this test by one question. It is critical that you use your scratch paper to draw as you solve the equation to answer the question.

MATH REVIEW

Rounding

Let's look at the concept of rounding. The numbers to the right of the digit, or point, begin with the tenth place. The second number is the hundredth place, and the third number is the thousandth place.

To round a number, begin by identifying the rounding digit, and look to its right side. If the digit to the right of your rounding digit is 4 or less, do not change it. All digits (numbers) to the right side of the rounding digit will be dropped.

Example: 0.3333   rounded to the nearest hundredth will be 0.33

What if the number to the right of your rounding digit is not 4 or less? If the digit is 5, 6, 7, 8, or 9, your rounding digit is increased by one number. All digits to the right side of the rounding digit will then be dropped.

Example: 0.6777   rounded to the nearest hundredth will be 0.68

Rounding to the nearest 100th

tenth
hundredth
thousandth

.376

5 ≥ = round second digit up
< 5 = second digit stays the same

Notes

For exam purposes, rounding is used most commonly when converting inches to feet (next lesson) for the purpose of multiplying and/or dividing like numbers. Since each of these steps will be written on a sheet of scratch paper, rounding to the nearest hundredth will save valuable time and will lead you to choose the correct answer.

If you do not round to the nearest hundredth, you will still arrive at the correct answer, but you are more likely to make a mistake.

Converting Inches to Feet

Converting inches to feet is necessary for multiplying like numbers and will frequently be the first step required to solve a math-related question. Begin to understand this concept by stating to yourself "36 inches is equal to 3 feet." Think about it. If someone asks how many feet are in 36 inches, your reply will be almost instant. This may be handy on test day.

Begin to think of inches in terms of fractions. To do this, 36 inches would be placed over 12 inches ($^{36}/_{12}$).

To convert a fraction to a decimal, you simply take the top number (numerator) and divide by the bottom number (denominator). Thirty-six inches divided by 12 equals 3. Apply this to any number presented as inches, and you can quickly convert it to feet.

Four inches converted to feet would simply be 4 divided by 12 ($^{4}/_{12}$). This equals 0.3333333333. Rounding to the nearest hundredth, 4 inches is equal to 0.33 feet.

Practice:

1. Rounding to the nearest hundredth, 14 inches is equal to ____ feet.

   a. 1.4

   b. 1.2

   c. 1.17

   d. 1.166

2. Rounding to the nearest hundredth, 8 inches is equal to ____ feet.

   a. 0.8

   b. 0.67

   c. 0.6666

   d. $^{8}/_{10}$

3. Rounding to the nearest hundredth, 3 inches is equal to ____ feet.

   a. 0.25

   b. 0.33

   c. 0.4

   d. 0.5

Answers/Solutions found in Part 17 of this book.

Applying the Concepts of Rounding and Converting Inches to Feet

The drawing shown represents the cross section of a two-story building. A *cross section* shows a view of the internal construction of an object or structure. It represents how it looks when cut on a plane.

The footing, slab, brick ledge, and building height can be determined from the drawing. A *footing* is a base or bottom of a foundation, pier, wall, or column that can be manually formed (stem wall) and supported by an earth-formed footing.

This particular footing will support a 4-inch brick and an 8-inch block separated by 2 inches of air. All heights for this building will be relative to the slab level and will be represented by an elevation number.

Notes

An *elevation* is the height above an established reference point, such as the slab level in the drawing. The elevation at the top of the slab is identified as 0. The level portion of the footing that will support the brick is identified as –67. This indicates the brick ledge is 67 hundredths of a foot lower than the floor level.

Answer the following questions based on the Wall Diagram on page 13.

1. Calculate the depth of the brick ledge. (D1(?))

    a. 0.6 ft.

    b. 0.67 ft.

    c. 0.8 ft.

    d. 0.85 ft.

2. What is the width of the footing? (D2(?))

    a. 1.2 ft.

    b. 1.17 ft.

    c. 1.4 ft.

    d. 1.25 ft.

3. What is the outside height of the footing? (D3(?))

    a. 2.1 ft.

    b. 1.2 ft.

    c. 1.75 ft.

    d. 1.9 ft.

4. What is the inside dimension, from grade to the top of the slab, for the footing? (D4(?))

    a. 2.9 ft.

    b. 2.42 ft.

    c. 2.35 ft.

    d. 2.25 ft.

Calculate Square Feet

$$\frac{L \times W}{2} \qquad \pi R^2 \qquad \frac{B + C}{2} \times H \qquad L \times W \text{ (or H)}$$

A majority of building components, including drywall, bundles of shingles, roof sheathing, brick, and block, are estimated by determining the square feet of the area first. Think of square feet as the foundation, or the starting point, for any take-off. If you understand how to properly calculate square feet, you can correctly answer many of the questions on the exam.

As discussed previously, you should be precise with your calculations. For example: If you are answering a question such as how many square feet of concrete slab will be installed, first determine the precise dimensions of the slab. Do not simply take the outside dimensions that have been provided. Read the details to determine where the slab is poured. Is it to the inside of the building, or does the building sit on top of the slab?

Example:

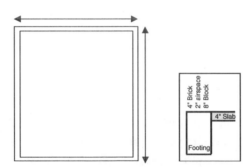

What is square footage of the 4-inch slab if the building dimension is 31 feet 6 inches × 20 feet?

- a. 632 sq. ft.
- b. 630 sq. ft.
- c. 583 sq. ft.
- d. 515 sq. ft.

Explanation:

The width of the footing must be subtracted from the overall dimensions of the building because the detail specifies that the slab is poured to the inside. The width is 14 inch, which converts to 1.17 feet.

> 20' − 1.17 − 1.17 = 17.66 (subtract the thickness twice)
> 31.5 − 1.17 − 1.17 = 29.16 (subtract the thickness twice)
> 17.66 × 29.16 = 514.9656 (answer is (d))

Notes

5. Based on the information provided, what is the dimension when measured on the inside of the structure?

    a. 14.67'

    b. 12.67'

    c. 13.5'

    d. 12.33'

6. If the construction of this building did not include brick and the dimensions of the interior remain the same as determined in Question #1, what is the measurement of the structure?

    a. 12.67'

    b. 13.67'

    c. 13.5'

    d. 11.67'

7. If this building were rectangular and measured 14 feet 6 inches in width and 23 feet in length, what is the inside dimension length-wise?

   a. 22.67'

   b. 22'

   c. 21.83'

   d. 20.66'

Using Area (Square Feet) to Estimate Sheathing, Drywall, Brick and Block

To estimate quantities of flat material, simply divide the area to be covered by the square feet of the material to be installed. For example, if the square feet of the interior walls for a structure totals 3,200, and 4' × 12' × ½" gypsum is being installed, divide 3,200 by 48 to determine that 67 boards must be ordered.

An obvious question at this point might be "Do I deduct for openings?" Do not deduct for openings when calculating gypsum/drywall. Do, however, deduct for openings when calculating brick or block.

1. How many sheets of 4' × 8' × ⅝" gypsum board will be needed for the interior of a garage with room dimensions of 24' × 30' with 8' ceilings? *Note:* The gypsum is to be installed on the walls and ceiling. Disregard openings.

   a. 23

   b. 27

   c. 50

   d. 69

2. After adjustments have been made to account for the pitch, a roof is determined to be 2,875 square feet. How many 4 × 8 pieces of sheathing will be installed?

   a. 90

   b. 60

   c. 120

   d. 52

3. Half-inch gypsum board is to be installed on the 8' walls of a room with dimensions of 12' × 14'. Specifications call for 4' × 8' × ⅝" gypsum board to be installed on the ceiling. The room has an opening of 6' on one side with two windows, each 4' × 6' facing the front of the structure. How many 4 × 12 sheets of ½" gypsum must be ordered?

a. 13

b. 9

c. 5

d. 4

Estimating brick and block is similar to estimating gypsum or sheathing. Begin by determining the face area of the material to be installed. For example, if an 8" × 16" nominal block (7⅝" × 15⅝" plus the mortar joint) is to be used, the face area of the block is 128 square inches. To convert this to a multiplier, you can follow one of several methods.

First, you can divide 144 square inches by 128 square inches to determine that it will take 1.125 blocks per square foot. The area (square feet) multiplied by 1.125 determines the number of 8" × 16" block needed.

You can also divide 128 square inches by 144 square inches to determine that one block will cover 0.8888 square feet. The area (square feet) divided by 0.8888 will determine the number of 8" × 16" block needed. Either way will result in the same answer.

Many contractors take the length of a wall and divide by 1.33 to determine the number of block required on each course. They will then take the height of the wall and divide by 0.67 (height of an 8 inch block) to determine the number of courses required. By multiplying the number of courses by the number of blocks on each course, the total quantity needed for the structure can be determined.

For test purposes, it is recommended that you follow one of the first two methods. Begin by determining the total area in square footage and multiplying/dividing by the appropriate factor.

Example:

How many blocks are needed for a wall that is 21 feet tall and 38 feet long?

    a. 898

    b. 600

    c. 917

    d. 1,191

Explanation:

Determine the area of the structure by multiplying the height times the length: $21' \times 38' = 898$ square feet. The area divided by 0.8888 will determine quantity of block needed: $798 \div 0.8888 = 898$ block. The area multiplied by 1.125 will also determine the quantity of block needed: $798 \times 1.125 = 898$ block. The answer is (a).

To estimate the quantity of brick needed for a structure, follow the same steps. However, we will look at only one way to determine the correct answer. Begin this task by determining the number of brick needed for each square foot of area. Next, determine the area, and multiply the area by the multiplier to arrive at the correct answer.

If a brick is $2\frac{1}{2}" \times 8"$, the face area of the unit is 20 square inches. The square inches of a square foot (144) divided by 20 equals 7.2 inches. If the area to be constructed with a brick veneer is 1,000 square feet, multiply the area by 7.2 to determine that 7,200 brick should be ordered for this project.

At this point, you may be thinking of waste. Factors will be provided on your exam to determine the allotted waste. If the waste factor on this project (question) is specified as 15%, simply add this amount to the total: 15% of 7,200 is 1,080; 7,200 plus 1,080 is 8,280. If this were a question on the exam, the correct answer would be 8,280 brick.

  4. Disregarding waste and loss due to overlapping at the corners, how many block would be needed for a 20' × 32' building that is 18' tall?

    a. 1,872

    b. 2,106

    c. 2,489

    d. 2,725

5. Based on a waste factor of 10%, how many brick will be needed for a 10' × 20' wall? The brick is 7" × 2¾".

   a. 1,295

   b. 1,485

   c. 1,646

   d. 1,760

6. A wall is 32' long, 8' high, and will be constructed out of 8" × 8" × 16" concrete masonry units (CMU) with a brick veneer. Total square feet of openings is 28, and the brick is 2½" × 8". How many of each is required?

   a. 1,642 brick, 257 block

   b. 1,487 brick, 228 block

   c. 1,287 brick, 188 block

   d. 1,189 brick, 171 block

Calculate square yards

$$\frac{L \times W}{9}$$

1 square foot

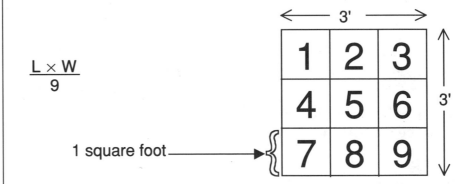

To calculate square yards, divide the area (square feet) by 9. There are 9 square feet in 1 square yard. It is highly recommended you always write the formula down before you begin to solve the equation. Write the formula down first, and carefully insert each number from the question into the formula.

It is not uncommon for an exam question to include too much infor-mation. In the testing profession, this information is referred to as a *distracter*. By paying attention to the formula, the distracter is less likely to cause a mistake in your calculations.

Example:

How many square yards of asphalt will be installed for a driveway that is 12 feet wide and 42 feet long? The depth of the asphalt is 2 inches.

    a. 85.68 sq. yd.

    b. 56 sq. yd.

    c. 112 sq. yd.

    d. 3.17 sq. yd.

Explanation:

Choices are provided for nearly any mistake made using the 2 inch depth. The proper way to calculate the square yards for this ques-tion is to calculate 12×42 and divide by 9. $12 \times 42 \div 9 = 56$. The answer is (b).

1. Calculate the square yards for a 20' $\times$ 45' area:

    a. 900 sq. yd.

    b. 33.33 sq. yd.

    c. 100 sq. yd.

    d. 75 sq. yd.

2. How many square yards of carpet must be ordered for a room that is 20' wide by 30' long? (This carpet only comes in a 12' roll.)

    a. 70 sq. yd.

    b. 80 sq. yd.

    c. 90 sq. yd.

    d. 100 sq. yd.

Notes

Notes

Calculate cubic yards

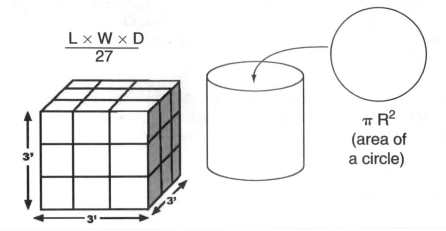

The formula to calculate cubic yards is length (in feet), times width (in feet), times depth (in feet), divided by 27. It is necessary to divide by 27 because there are 27 cubic feet in 1 cubic yard.

Cubic yards will often be referred to as *volume*. If an exam question directs you to calculate the volume of a building component, it is most likely asking for the cubic yards. However, if the choices are provided in cubic feet, do not divide by 27 (CF = L × W × D). Concrete, sand, and dirt are most commonly estimated in volume of cubic yards.

To calculate the cubic yards of a cylinder, begin by computing the area of the base. Multiply the base by the height of the cylinder and divide the total by 27.

Example:

Calculate the cubic yards of concrete needed for a 6 foot column that is 30 feet tall.

    a. 20.93 cu. yd.

    b. 31.4 cu. yd.

    c. 565.2 cu. yd.

    d. 847.8 cu. yd.

Explanation:

Begin with the formula L × W × D ÷ 27. L × W is the area of the column base computed by πR^2. R = half of the diameter, which will be 3, square the 3 to get 9. Pi × 9 equals the area, so 3.14 × 9 gives the area to be 28.26. 28.26 is the L × W portion of the CY formula. Now, multiply this by the depth of 30 feet to determine that this column has 847.8 cubic feet. Divide the cubic feet by 27 to determine the cubic yards: 847.8 ÷ 27 = 31.4. The answer is (b). (Note π is symbol for pi which always equals 3.14)

Example

How many cubic yards of concrete should be ordered for a 4"-deep drive that is 14' wide and 65' long?

 a. 8.43 cu. yd.

 b. 11.12 cu. yd.

 c. 300.3 cu. yd.

 d. 364.7 cu. yd.

Explanation:

Begin with the formula L × W × D ÷ 27. The length is 65, the width is 14, and the depth is 0.33 (4 divided by 12): 65 × 14 × 0.33 ÷ 27 = 11.122. The answer is (b).

1. How many cubic yards of concrete will be needed for a slab to be poured for a parking garage if the surface area is 875 square feet and the average slab depth is 6 inches?

 a. 10.80 cu. yd.

 b. 16.20 cu. yd.

 c. 19.44 cu. yd.

 d. 437.5 cu. yd.

2. Calculate the volume for a 32' tall holding tank that is 18" in diameter.

 a. 254.34

 b. 8,138.88

 c. 301.44

 d. 354.64

3. Two hundred linear feet of footing will be poured with an average width of 14". The average depth of the footing is 16". Concrete must be ordered by the whole yard and costs $65 per unit. What is the cost of the concrete for the footings?

 a. $715.00

 b. $749.45

 c. $780.00

 d. $2,022.93

Notes

4. Based on this detail, calculate the cubic yards of sand needed to place between the slab and grade. The interior dimensions of the structure are 14' × 29'.

   a. 678 cu. yd.

   b. 25.11 cu. yd.

   c. 19.99 cu. yd.

   d. 18.65 cu. yd.

Calculating cubic yards is simple, especially when you are asked to calculate concrete. If the question is in reference to cubic yards of dirt, the swell and compaction factors need to be considered.

Swell is defined as the volume growth in soil after it is excavated. If you excavate, the soil will expand as it loosens. You cannot predict the number of dump trucks necessary to haul the excavated material away based solely on the dimensions of the area to be excavated. The volume of soil at its natural state, or bank cubic yards, will exceed the area excavated because it is no longer compacted.

The percentage of swell, or loose soil, depends on the soil type. This will be specified on the exam unless one of your approved reference books includes a chart with the information. If you are asked to calculate the cubic yards of dirt to be excavated from an area and are provided with a swell factor, simply increase the volume of the area to be excavated by the percentage provided.

Compaction rates can also be specified. When an excavation is to be backfilled, the fill must often be mechanically compacted. The compaction factors, or the difference between the fill in the truck and the final in-place compacted quantity, must be considered. If the compaction factor is 15%, divide the volume of the excavation by the reciprocal of the compaction factor. For example, if the open excavation represents 9 cubic yards and the compaction factor is 15%, divide 9 by 0.85. The volume to be delivered to fill the excavation after compaction is 10.58 cubic yards.

Example:

Based on a swell factor of 20%, how many cubic yards of dirt will be hauled away from a 12' × 22' excavation measuring 6' in depth?

   a. 1,900.8 cu. yd.

   b. 58.67 cu. yd.

   c. 73.33 cu. yd.

   d. 70.39 cu. yd.

Explanation:

Begin with the formula L × W × D ÷ 27. The length is 12, the width is 22, and the depth is 6. 12 × 22 × 6 ÷ 27 = 58.67 plus 20% = 70.399. The answer is (d).

Example:

Based on a compaction loss of 10%, how much fill is needed for an open excavation with a volume of 21 cubic yards?

    a. 23.1 cu. yd.

    b. 30 cu. yd.

    c. 23.33 cu. yd.

    d. 26.25 cu. yd.

Explanation:

21 ÷ 0.90 = 23.33. The answer is (c).

1. Based on Drawing #1, determine the quantity of soil to be hauled away from this site if it needs to be excavated 3' deep. Include an average of 2' around the entire structure. The swell factor is 15%, and the grade is considered to be level.

    a. 244.3 cu. yd.

    b. 265.5 cu. yd.

    c. 277.3 cu. yd.

    d. 280.6 cu. yd.

2. Soil with a 20% swell factor will be removed from a 4,000-square-foot excavation that is 7 feet deep. A dump truck will haul 20 cubic yards. How many dump trucks will be needed for this project?

    a. 52

    b. 63

    c. 67

    d. 71

Notes

Drawing # 1

Notes

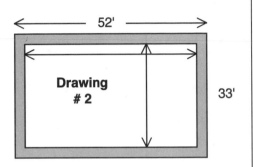

Drawing # 2

52'

33'

3. Considering a 10% loss of compaction, how many cubic yards of dirt will be needed to backfill an area that is 14' × 72' with a depth of 4'?

   a. 149.33 cu. yd.

   b. 164.266 cu. yd.

   c. 165.925 cu. yd.

   d. 167.854 cu. yd.

4. Based on Drawing #2, this area has been excavated 5' from the exterior of each wall. The entire area is 2' too low and must be backfilled with engineered soil that has a loss of compaction rated at 15%. How many cubic yards should be ordered?

   a. 232.33 cu. yd.

   b. 245.55 cu. yd.

   c. 254.32 cu. yd.

   d. 257.31 cu. yd.

FLOOR FRAMING TERMINOLOGY

Decking Material that forms the floor surface, usually attached directly over the floor joists

Girder A larger beam of wood or steel used as the main support for concentrated loads at points along its span

Floor joists Parallel framing members to support floors. These are the main framing members that support the floor span. Joists are usually made of engineered wood I-beams or at least 2 × 8 lumber.

Joist hangers Brackets designed to hold joist ends

Ledger strip A strip of lumber nailed to a beam, girder, or joist on which the floor joist rests for support

On center The allotted spacing between studs, joists, and rafters. This measurement is taken from the center of one member to the center of the adjacent member. On-center spacing is dictated by code and effects the joist size.

Span Distance between the structural supports in floors, ceilings, and roofs.

WALL FRAMING TERMINOLOGY

header   top plate

Nominal VS Dressed
(Name Size VS Actual Size)

stud

Actual dimensions,
upto 6" - subtract 1/2"
more than 6" - subtract 3/4"

a 2 × 4 is actually 1½" × 3½"
a 2 × 8 is actually 1½" × 7¼"

*Code allowances are based
on actual dimensions.*

king stud
trimmer (jack)

cripple

sole plate

16" On-Center Spacing

16"   16"   16"

Stud Count = L÷ O.C. Spacing +1

Bottom (sole) plate The bottom horizontal structural member of a stud-framed wall. The bottom plate is fastened to the floor (foundation or subfloor) with foundation bolts or proper nails.

Cripple stud The short studs placed above or below an opening. Cripples attach the header to the top plate and the sill to the sole plate.

Header A beam placed perpendicular to the wall studs above openings. The header is sized based on the span and the weight to be supported by the member.

Rough opening The framed opening to accommodate future installation of a door, window or open passageway. Measurements for placing such components are made to the center of the rough opening.

Stud A 2" × 4" or 2" × 6" vertical framing member used to construct walls and partitions.

Top plate The top horizontal framing member that caps the framed wall and makes the connection for the ceiling joists and/or rafters. Top plates are usually doubled on exterior and bearing walls and can be single members for interior nonbearing partitions.

Trimmer stud The vertical member, sometimes referred to as a jack, that supports a header.

King stud The vertical framing members on the outside of the jack or trimmer stud.

WALL FRAMING MATH

Let's begin this topic with a review of the terminology used for estimating lumber. As reviewed previously, *square feet* is the area determined by multiplying length times width. For estimating lumber, it is how much surface a milled wood product will cover.

Another common unit of measuring for construction is *linear feet*. This unit of measurement is one dimensional. It is the measurement of the framing member's length or an aggregate measurement of the entire wall length. It is common to begin a take-off (estimate) by determining the total linear feet of walls to be constructed.

Board feet, a less common unit of measurement for construction, is three dimensional and measures the volume of wood when sawn before it is kiln-dried. It is important to note that wood shrinks, more in width than in length, so the proper volume of lumber must be available before the desired finish product is planed. To estimate the quantity of board feet, multiply the quantity (number of pieces) by the nominal size of the member (in inches) by the length of the member (in feet) and divide by 12. Yes, you will break the rule of always multiplying "like numbers." This is the proper way to estimate board feet.

Nominal size is the size of the lumber before it is planed or finished. It is the "name size." The actual dimension of the framing members is not the same as the name size. The actual size of the framing members is reduced by ½" for nominal sizes up to 6" and ¾" for members above 6". For example, a 2 × 4 is actually 1½" × 3½". A 2 × 8 is actually 1½" × 7¼". Understanding this distinction is important to properly apply the building code and is critical to accurately performing math operations on the exam.

On-center spacing is an important concept to understand when estimating framing materials. If 16" on center is specified, the member will straddle the 16" incremental point across the span. The first two members are not center to center. Initially, the tape is pulled from a corner, and marks are made at the first 16-inch point and from each one that follows. Proper centering is important for the purpose of sheathing installation.

Dimensions on a set of plans are provided based on center to center of the framing. Construction materials often vary slightly in size, making it impossible to indicate actual edges of structural members. This method of measurement ensures that the plan (construction document) dimensions are achieved. If a window is to be placed, measurements are made to the center of the rough opening. The jack is installed by measuring a distance in each direction away from the center point that is equal to half of the rough opening. The size of the rough opening is always as wide as the specified component size plus 2 to 3 inches. The additional space provides for shimming the component to be installed so that it is level and plumb.

The quantity of members necessary for proper construction, such as studs, is calculated by taking the linear feet of the span and dividing it by the on-center spacing. This results in the number of spaces between the units; therefore, one additional unit must be added for the beginning (first) member. For example, if 2 × 4 studs are to be installed 16" on center in a 32' wall, the quantity of studs are calculated by taking 384" (32' × 12 to convert to "like" numbers) and dividing by 16. The result is 24. There are 24 spaces between each of the studs, add 1 and you have identified the number of studs to be installed in the wall. This same method applies to calculating the necessary number of joists or rafters.

If this 32' wall is 8' tall including a single sole plate, a double top plate, and no openings, the linear feet of lumber can easily be calculated. Begin by subtracting the thickness of the plates from the height of the wall. Since 2 × 4's are being installed, the thickness of each plate is 1½" (nominal vs. actual size). Three plates at 1½" each totals 4½". Subtract 4½" from 96" to determine that each stud will be 91½" in length.

If 25 studs are to be installed for this 32' wall, 25 × 91.5 gives us a total of 2287.5". Add the length of the plates, three plates at 32' long for a total of 96'. Now, divide 2287.5" by 12 to convert to feet. The total linear feet of lumber of studs is 190.62½. Add 96' for the plates; there are 286.62½ linear feet of lumber in this wall.

While the linear feet of lumber is rarely used in the real world, understanding the concept is critical to accurate estimating. More importantly, at this point, the concept is common on exams.

To calculate the board feet of lumber for this wall, begin by multiplying 25 × 2 × 4 × 8 and dividing by 12. This equals 133.33. This is the total number of board feet in the studs. Next, multiply 3 (the number of plates) by 2 × 4 × 32 and divide by 12. This equals 64. Add the two totals, 170.66 and 64; the total board feet for this wall is 234.66. When calculating board feet, you do not deduct for the material that will be cut, as we did when calculating the linear feet.

To "lay out a house" for the framer, it is necessary and critical to have a thorough understanding of plan reading and the ability to

Notes

Notes

work a calculator. For exam purposes, the understanding is even more critical because you will have to visualize and sketch the exercise. For example, in the real world, it would be simple to locate the position of a 3' door to be installed 10' from the edge of the room. The plans would indicate that the 10' is the center of the 3' door. A seasoned carpenter would simply pull the measuring tape from the edge of the room and mark the 10' point. He would then add 2½" to the size of the door and divide the total by 2. This would provide the second and third pencil mark to be made. Half of 38½" is 19¼". The carpenter would measure 19¼" to the right of the 10' mark and 19¼" to the left of the 10' mark. These marks would indicate where the edge of the trimmer stud (jack) should be installed. The following examples indicate how this question may be presented on an exam.

Example:

The center of the rough opening for a 3' door is indicated on the plans as being 10' from the edge of the room. What is the measurement from the edge of the room to the first beginning point of the rough opening?

    a. 100.75"

    b. 99.25"

    c. 98.75"

    d. 97.25"

Explanation:

Begin by sketching the information as it is described. Make a dimension line from the left to right indicating 10' to the center of an opening. The opening is 38.5 inches —the 3' for the door plus 2½" for the rough opening. Divide the rough opening by 2. Draw the dimension line, indicating 19.25". Subtract 19.25" from 120": 120 − 19.25 = 100.75". This is the measurement from edge of the room to where the rough opening begins. The answer is (a).

Example:

The center of the rough opening for a 3' door is indicated on the plans as being 10' from the edge of the room. What is the measurement from the edge of the room to the first right edge of the first trimmer?

    a. 100.75"

    b. 99.25"

    c. 98.75"

    d. 97.25"

Explanation:

Begin by sketching the information as it is described. Make a dimension line from the left to right indicating 10' to the center of an opening. The question is asking for the measurement to the right edge of the first trimmer. The header rests on the trimmer; therefore, the measurement to be identified will also indicate the measurement to the beginning of the header. The measurement of the header is 3" larger than the rough opening. This is determined by adding the width of each trimmer used to hold the header in place. Each trimmer is 1½" for a total of 3". The header is 41½". Divide 41.5 by 2 to determine half of the header. Subtract 20.75 from 120". The measurement from the left to the first right edge of the trimmer is 99.25". The answer is (b).

1. How many studs will be needed for a wall that is 25' long if studs are placed 16" on center?

    a. 18

    b. 19

    c. 20

    d. 21

2. Calculate the board feet of lumber in 18 floor joists. The 2 × 10 floor joists are 16' long.

    a. 27

    b. 480

    c. 1,280

    d. 5,760

Notes

3. What is the measurement to the beginning of the rough opening for a 3' door? The door will be placed with the center located 9' from the edge of the room?

   a. 86.75"

   b. 87.25"

   c. 88.75"

   d. 89.25"

4. How many floor joists will be needed for a 24' span if spaced 16" on center?

   a. 18

   b. 19

   c. 20

   d. 24

ROOF FRAMING TERMINOLOGY

Ridgeboard The highest horizontal roof member which serves to align the rafters and tie them together at the upper end. The ridgeboard is at least one size larger than the rafters.

Common rafter A structural member that extends from the top plate to the ridge in a perpendicular orientation. Rafters often extend beyond the roof plate to form the overhang, also called the eaves, which protect the sides of the structure.

Bird's mouth The notch, or cutout, of the rafter, which allows it to rest properly on the top plate.

Hip rafter A roof member that extends diagonally from the corner of the plate to the ridge.

Valley rafter A roof member that extends from the plate to the ridge along the lines where two roofs intersect.

Jack and hip rafters These types of rafters do not extend the entire distance from the ridge to the top plate of a wall.

Cripple jack A rafter fitted between a hip rafter and a valley rafter, a cripple jack does not touch the ridgeboard or top plate.

ROOF TYPES

Gable roof Perhaps the most common, a gable roof has two slopes that meet at the center of the building. It is simple, economical, and can be used on virtually any structure.

Hip roof A hip roof has four sides or slopes running toward the center of the building. Rafters at the corners extend diagonally to meet at the ridge. Additional rafters are framed into these rafters.

Gable and Valley roof A roof on which two gable roofs intersect.

Hip and Valley roof A roof on which two hip roofs intersect.

Mansard roof A roof that has four sloping sides, each of which as a double slope. As compared with a gable roof, this design provides more available space in the per level of the building.

Gambrel roofing A variation on the gable roof in which each side has a break near the ridge. This style of roof will provide more available space in the upper level.

Shed roof Also referred to as a lean-to roof, the shed roof is a flat, sloped construction. It is common on high-ceiling contemporary construction and is often used on additions.

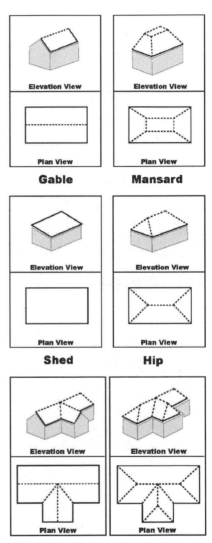

Gable   Mansard

Shed   Hip

Gable & Valley   Hip & Valley

Gambrel

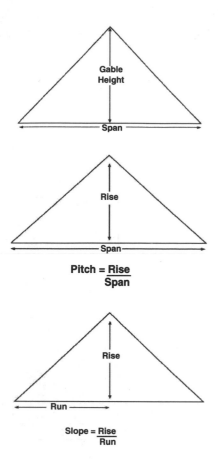

Gable Height
Span

Rise
Span

$$Pitch = \frac{Rise}{Span}$$

Rise
Run

$$Slope = \frac{Rise}{Run}$$

ROOFING MATH

To begin the discussion of roofing math, there are several terms that must be clearly understood. It is important to understand that the term *span* refers to the horizontal distance from the outside of one exterior wall to the outside of the opposite wall. *Run* is the horizontal distance from the outside of the top plate to the center line of the ridgeboard, usually equal to half the span.

Rise is the total height of the rafter from the top plate to the ridge. The rise is typically stated in inches per foot of run.

Pitch is the angle or degree of slope of the roof in relation to the span. It is a ratio of gable height over the full span of the roof. Pitch is expressed as a fraction. For example, if the total rise is 6' and the span is 24', the pitch would be 6 in 24 which reduces to a ¼ pitch.

Slope is expressed as the relationship of rise to run. It is stated as a unit of rise to horizontal units of 12. A roof that has a rise of 5" for each foot of run is said to have a 5 in 12 slope.

It is common for a seasoned contractor to refer to a 5 in 12 slope as a $\frac{5}{12}$ pitch. No one corrects him and asks, "are you sure you don't mean a 5 in 12 slope?" Notice, the only words that change between the two statements are *pitch* and *slope*. These terms are used interchangeably, and both uses are accepted as correct. In fact, it is so common that even the codebooks refer to a slope as a pitch. Let's look at some examples to make sure that this concept is understood.

If a building is 24 feet wide and has a gable height of 8 feet, by placing the height over the building span, the pitch will be 8 over 24. This fraction reduces to ⅓.

If you have forgotten how to reduce a fraction, simply divide the numerator (top number of the fraction) and the denominator (bottom number) by their greatest common factor or devisor. If you have properly reduced the fraction, only the number 1 can be divided evenly into both the final numerator and denominator.

The slope of this roof would be 8 in 12, because a slope is expressed in a ratio of rise over run. Half of the 24' span would be a 12' run. The rise of 8 and the run of 12 is expressed in a ratio format of 8 in 12.

Let's try another example: If the building is 32' wide with a gable height of 8', by placing the numbers in a ratio of gable height over the span of the building, you will have 8 over 32. Now, reduce the fraction by taking the greatest common factor and dividing it into the numerator and the denominator. The greatest common factor is 8, and 8 will go into 8 one time, and 8, will go into 32 four times, with the result being that this roof has a one-fourth pitch.

If you need to determine the slope of this roof, converting the ¼ pitch to a slope is a little more difficult. Roof slopes are always provided with a ratio of the specified units of rise to 12 units of horizontal run. The run is always indicated in units of 12. This is because we measure in units of 12". When asked, a contractor usually provides measurements in feet. Rarely will he say "the building is 480" wide."

So, what is the slope of a roof 32' wide with an 8' gable height? Many would quickly say 8 over 16. Yes, but this would not be one of the choices on the exam. The choices would be in a ratio of rise to run with the run being 12. So, let's set it up to solve for *x* (or the correct unit of rise).

Place *x* over 12, and beside it put 8 over 16. We will now cross multiply and divide by the reciprocal to solve for *x*: $12 \times 8 = 96$. Divide 96 by 16, which equals 6. The slope of this roof is 6 in 12.

Step 1: $\dfrac{x}{12}$ $\dfrac{8}{16}$

Step 2: $\dfrac{x}{12}$ ↗ $\dfrac{8}{16}$

(cross multiply)
$12 \times 8 = 96$

Step 3:   Divide by reciprocal:
$96 \div 16 = 6$

Example:

Determine the slope of a roof in units of vertical rise to 12 units of horizontal run when the span of the roof is 30' and the height of the gable is 10'. By placing the rise over the run, you have 10 over 15. However, we need it in units of rise to units of horizontal run. Again, we will solve

Notes

for x over 12 and beside it write the fraction 10 over 15. We will cross multiply, so 12 times 10 will equal 120. Now, divide by the reciprocal of 15 to get 8. The slope of this roof is 8 in 12.

Step 1:	$\dfrac{x}{12}$	$\dfrac{10}{15}$
Step 2:	$\dfrac{x}{12}$	$\dfrac{10}{15}$
	(cross multiply)	
	$12 \times 10 = 120$	
Step 3:	Divide by reciprocal:	
	$120 \div 15 = 8$	

If provided a pitch in its reduced fraction such as ⅓, follow the same steps. The slope will be 1 over 1.5 (If 3 equals the span, run is half hence the 1.5.) To convert this to a slope, place x over 12 beside the fraction 1 over 1.5. Cross multiply 12 times 1 and divide by the reciprocal of 1.5. The result is 8. A roof with a ⅓ pitch will have an 8 in 12 slope.

Step 1:	$\dfrac{x}{12}$	$\dfrac{1}{1.5}$
Step 2:	$\dfrac{x}{12}$ ↗	$\dfrac{1}{1.5}$
	(cross multiply)	
	$12 \times 1 = 12$	
Step 3:	Divide by reciprocal:	
	$12 \div 1.5 = 8$	

Questions on this topic generally use whole numbers and reduced pitches. It may be to your advantage to make a chart with three columns. The first column will be labeled "slope," the second column will be labeled "pitch," and the third column will be labeled "reduced pitch." First, fill in the slopes beginning with 2 in 12. Continue down to 8 in 12. In the second column, convert the pitch to a slope by multiplying 12 by 2 and placing the vertical rise on top (see example). You have converted the run to a span and now have a ratio of rise over span. In the last column, reduce each of the fractions.

A typical exam question is What is the vertical rise of a roof that has a ⅓ pitch?" Using your chart, locate the ⅓ pitch, follow it to the column labeled "slope," and identify the vertical rise to be 8.

Run

$\dfrac{\text{Rise}}{2 \times \text{Run}}$

Slope	Pitch	Reduced Pitch
②:12	2/24	1/12
3:12	3/24	1/8
4:12		
5:12		
6:12		
7:12		
8:12		
9:12		
10:12		

Rise

Example:

What is the slope of a gable roof with a 36' span? The height of the gable is 12'.

    a. 2:12

    b. 3:12

    c. 4:12

    d. 5:12

Explanation:

Place *x* over 12, and beside it write 12 over 36. Cross multiply: $12 \times 12 = 144$. Divide by 36. $144 \div 36 = 4$. The answer is (c).

Gable Height

Determining the gable height of a structure is simple as long as the roof slope and span are provided. If the span of a building is 12' and the slope is 8 in 12, take half of the span and multiply it by 8". This formula is confusing to some people because it seems to be multiplying feet by inches (breaking the rule to always multiply "like numbers"). Do not think of the 6 as feet. Consider the 6 to be units and tell yourself that for each of the 6 units (units in 12-inch increments), the roof rises 8". (See diagram below)

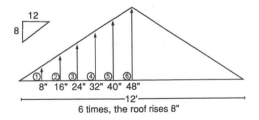

For a building with a span of 24' and a roof sloped 4 in 12, first determine the measurement to the ridge. If the building is 24' wide, the ridge will be located in the center. Half of 24 is 12. Now, multiply 12 by the vertical rise for a total of 48 inches. Divide 48 by 12, and the gable height is determined to be 4 feet.

Understanding this concept allows you to easily calculate the area (square feet) of the gable. Once it is calculated, the area can be multiplied or divided by the appropriate factor to estimate siding, brick, sheathing, and similar building components.

Notes

If you need to estimate the quantity of brick (based on a factor of 7) to be installed in the gable area of a building that is 30' wide with an 8 in 12 slope roof, begin by converting the 30' span to a run. Divide 30 by 2. The run of the building is 15'. Next, multiply 15 by the vertical rise of 8. This equals 120. The height of the gable is 120" or 10'. Determine the area of the gable by taking the length of the run and multiplying it by the height of the gable. The area of the gable is 150 square feet. Since the actual area is only half of the rectangle that you calculated, it will account for each portion of the gable area. Take the square feet of the gable, and multiply it by 7 to determine that 1050 brick will be needed for this area.

Example:

Calculate the square feet of a gable for a building that is 28' wide and built with a 4:12 slope.

    a. 32.69 sq. ft.

    b. 65.38 sq. ft.

    c. 78.36 sq. ft.

    d. 95.25 sq. ft.

Explanation:

Identify the center point of the span, which is the highest point of the gable: 28 ÷ 2 = 14. This roof rises 4" for each foot. Multiply 14 by 4 to determine the gable height: 14 × 4 = 56. Convert 56 inches to feet: 56 ÷ 12 = 4.67'. Multiply 14 by 4.67 for a total of 65.38 square feet. If calculating the area of a triangle, divide by 2. It is not necessary to divide by 2 in this case because you have two triangles, one on each side of the ridge. The answer is (b).

1. What is the gable height for a building that is 28' wide? The roof slope is 6 in 12.

    a. 7 ft.

    b. 8 ft.

    c. 9 ft.

    d. 10 ft.

2. What is the slope of a roof with a ¼ pitch?

    a. 4:12

    b. 5:12

    c. 6:12

    d. 8:12

3. What is the run of a roof that has a 29' span?

    a. 13 ft.

    b. 13.5 ft.

    c. 14 ft.

    d. 14.5 ft.

4. Calculate the area of a gable if the slope of the roof is 8:12. The width of the building is 32'.

    a. 256.00

    b. 170.72

    c. 225.80

    d. 154.75

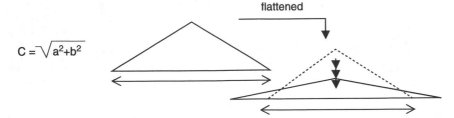

$$C = \sqrt{a^2 + b^2}$$

The next concept is critical for many areas of roofing math, including estimating rafter lengths, bundles of shingles, and rolls of felt. For explanatory purposes, suppose we are working with a gable roof that is 24' wide with a 4 in 12 slope. Imagine that the ridge-board is hinged to allow the roof to lay flat. If this roof were laid flat, it would not be 24' wide. Obviously, it would be wider. The question is, how much wider?

To determine the measurement of a building if we could lay it flat, we must apply the Pythagorean theorem. This theory states that when we have a right triangle, one that is formed with a 90-degree angle at the base, we can determine the hypotenuse as long as we

Notes

know the two remaining measurements. Speaking in construction terms rather than mathematical jargon like triangles and hypotenuses, the right triangle is one half of our 24' wide building. The hypotenuse is the rafter on the building that connects the base to the ridge.

The formula states that C, the rafter, is equal to the square root of A squared plus B squared. A is the run of the building, and B is the rise of the building. In this case, A is 12, and B is 4. Plug the numbers into the formula. Begin by multiplying 12 by 12. This equals 144. Next, multiply 4 by 4. This equals 16. Add the two together; 144 plus 16 equals 160. Find the square root symbol on your calculator, and press it. The numbers returned are 12.6491. This tells us the length of the rafter for this building. Double this number, and the dimension of the roof is 25.2982. If we were answering a question on the exam, it would likely have asked for the rafter length to be ordered. Since lumber is ordered in 2' increments, the correct choice for this question is 14'.

The 24' wide building with a sloped roof of 4 in 12 is the most simple of this type of mathematical operation. If the width of the building is 30' and the slope remains 4 in 12, another step must be included in solving the equation. It is important to realize why this is so. In the example of the 24' wide building, the height of the gable happened to be the same as the rise in the 4 in 12 slope. If the building were 30' wide with a 4 in 12 slope, the gable height would be different. The B in the formula stated in the Pythagorean theorem as we used it represents the gable height. So, if the building is 30' wide, begin by calculating the height of the gable. The center point of the building (peak of the gable) is 15'. If the slope is 4 in 12, we must multiply 15 by 4. This tells us that the height of the gable is 60". To convert 60" to feet, divide by 12. The height of the gable is 5'.

We can now calculate the length of the rafter (hypotenuse) of the building (half of which represents a right triangle). A is 15'. Fifteen squared equals 225. B is 5'. Five squared equals 25. Add 225 and 25, for a total of 250. Enter 250 in the calculator, and press the square root symbol $\sqrt{\ }$. The calculator returns 15.8113. The material to be ordered for the rafters in this building will be 16' in length.

Example:

A gable roof is built with an 8:12 slope. The building is 32' wide. If the two rafters that meet at the ridge are laid end to end, what is the total linear feet?

    a. 19.23 linear feet

    b. 34.25 linear feet

    c. 36.84 linear feet

    d. 38.46 linear feet

Explanation:

First, determine the gable height, take half of 32, which identifies the location of the highest point of the gable, and multiply 16 by 8 (this roof rises 8" for each foot): $16 \times 8 = 128"$. Convert to feet: $128 \div 12 = 10.67'$ (gable height). Now, plug the numbers into the Pythagorean theorem. $A = 16, B = 10.67, 16 \times 16 = 256, 10.67 \times 10.67 = 113.85$. Add these together: $256 + 113.85 = 369.85$. Press the square root symbol to get 19.23. $19.23 \times 2 = 38.46$. The two rafters end to end are 38.46'.

If the roof multiplier for a 4 in 12 slope roof had been available for the 30' wide building, the calculation would have been much simpler. If we were calculating the area of the roof, we would have multiplied the 30' by the multiplier. If we were calculating the length of a rafter, we would have divided the 30 by 2 to determine the run and multiplied the run by the factor, which would give us the rafter length. The step for determining the gable height would have been eliminated.

The roof multiplier for a 4 in 12 slope roof is 1.054. How is this calculated? Simply, plug the numbers into the Pythagorean theorem, and divide the answer by 12. This indicates the degree that the rafter (hypotenuse) rises per foot of run.

Let's calculate the roof factor for a 6 in 12 sloped roof. Plug the numbers into the formula for Pythagorean theorem, C equals the square root of A squared plus B squared. Twelve squared equals 144; 6 squared equals 36; 144 plus 36 equals 180. Press the square root symbol on the calculator $\sqrt{\ }$, and 13.4164 is returned. Finally, divide this by 12. Carry the factor out to the third digit. The roof multiplier, or factor, is 1.118.

Notes

Notes

It is not uncommon for the exam to provide the roof multiplier in the instructions for the related math question. Of course, if you do not realize how to use the factor, it does you no good. If the factor is not provided, it can be easily be calculated.

Calculate the factors for each of the slopes and fill in the table:

Slope	Equation	Multiplier
2:12	$2 \times 2 = 4$, $12 \times 12 = 144$, $4 + 144 = 148$, press $\sqrt{\ } = 12.1655$, divide by 12 = 1.013.	**1.013**
3:12		
4:12		
5:12		
6:12		
8:12		

5. Calculate the length of a common rafter for a 20' × 50' gable roof. The roof slope is 4:12, and the building has a 1' overhang.

  a. 10 ft.

  b. 12 ft.

  c. 13 ft.

  d. 22 ft.

6. If 6:12 gable roof is 30' wide with a 1' overhang, what is the measurement to use in calculating the roof area?

  a. 33.54

  b. 34.658

  c. 35.776

  d. 36.516

7. Using the Pythagorean theorem, what is the rafter length for a building that is 27' wide and measures 6' at the gable height?

    a. 12.56 ft.

    b. 14.77 ft.

    c. 15.23 ft.

    d. 16.85 ft.

Roof area can easily be calculated using the multiplier, allowing us to compute the bundles of shingles, rolls of felt, or pieces of sheathing needed for the roof. You can simply calculate the square feet of the home and increase it by the roof multiplier. You may be asking, "What if it is a hip roof as opposed to a gable roof?" Believe it or not, it matters so little that it is acceptable practice to estimate both using this method.

Can you simply take the square feet provided as square feet for the roof area? You could, but if the structure has an overhang, it has not been calculated in the finished square feet. The overhang is the portion of the roof that extends beyond the wall of the building. There is no standard overhang. In fact, overhangs vary based on different roof pitches, architectural styles, and personal preferences. A structure can have multiple overhangs and be different for gable ends than what is specified for the eaves.

The simple way to begin calculating the area of the roof is to sketch the plan view of the structure. This is simply a two-dimensional sketch of the footprint of the building as if you were looking down on top of it. Next, you should place a dotted line around the perimeter of your sketch to illustrate the overhang. The area of the roof should be calculated using the new dimensions. This will provide the area of the roof if it were flat. Once this calculation is achieved, you simply multiply this number by the roof multiplier to account for the rise created by the pitch.

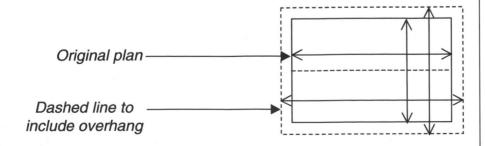

Original plan

Dashed line to include overhang

Notes

Let's try a simple roof. Calculate the roof area of a building that is 30' wide, 45' long, with a roof slope of 5:12. The overhang for the building is 1 foot. Begin by sketching the 30 × 45 foot rectangle. Next, draw a dotted line around the perimeter sketch to indicate the overhang. Now, draw a dimension line to connect the dashed lines on each side, and write the new dimensions to include the original measure plus the overhang.

The new dimension will be 30' plus 1' of overhang on each end, for a total of 32'. The dimension line going in the opposite direction will be 45' plus 1' of overhang on each end, for a total of 47'. Multiply 47 by 32 to determine the square feet of the flat area of this structure including the overhangs. Forty-seven feet times 32' equals 1,504 square feet. Finally, multiply the 1,504 square feet by 1.083. This is the multiplier for a 5 in 12 sloped roof. The total square feet of the roof for this building is 1,628 square feet.

Let's try another example. Calculate the area of a hip roof that is 75' in length, 27' in width, and has an 18" overhang. The slope of the roof is 6 in 12. Sketch the footprint of the building. Next, draw a dashed line around the perimeter of the building sketch to indicate the overhang. Add dimension lines to connect the dashed lines that are parallel to one another. The new dimensions will be 78' × 30'. Multiply 78 × 30; the area of the roof is 2,340 square feet. Multiply 2,340 square feet × 1.118, the multiplier for a 6:12 slope, to determine that the total area of the roof is 2,616 square feet.

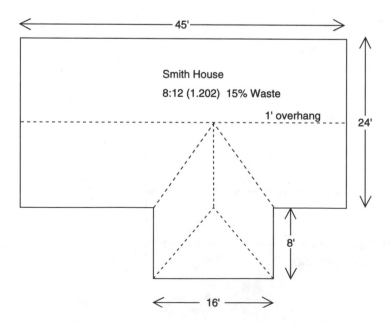

Now, one that is a little more difficult (see diagram on previous page). The building in the diagram provides all of the information necessary for calculating the area of the roof. It provides the roof slope as well as the roof multiplier. The waste factor for this plan is indicated, and the overhang is noted. Follow these steps:

In Step 1, the dashed line is added to indicate the overhang, In Step 2, the new dimension lines were added to show the overhang. Step 3 divided the plan into manageable sections. The larger of the two sections measures 47' × 26'. The smaller section is 18' × 8'. Notice that the 8-foot dimension line for the small section stays the same because of the one foot included in the larger section (indicated by the heavy line).

The larger section is 1,222 square feet, and the smaller section is 144 square feet. Added together, the total for the roof area is 1,366 square feet. To account for the pitch, multiply 1,366 × 1.202, which equals 1,641.93. Add 15% to consider the waste. The total area of roofing for this plan is 1888.22 square feet.

If you were asked to calculate the bundles of 15#, three-tab shingles for this plan, you would divide the total area by 100. This is to convert the roof area to squares. There is 100 square feet per roofing square. Shingles are packaged such that three bundles will cover 100 square feet. By multiplying the squares by 3, you determine that this plan will require 57 bundles of shingles (1,888.22 ÷ 100 = 18.88 × 3 = 56.64, which rounds to 57 bundles).

Questions on the exam typically include the roof multiplier, considerations for waste, the number of bundles required per square, yield for felt, and a statement to disregard starter shingles and ridge caps.

Notes

Step 1

Sketch dash line to indicate overhang.

Step 2

Add dimension lines to include overhang.

Step 3

Divide the plan into sections.

Notes

8. Calculate the square feet of this building. *Note:* Roof slope is 4:12, overhang is 1', waste factor is 10%.

    a. 2,570 square feet

    b. 2,824 square feet

    c. 2,976 square feet

    d. 3,274 square feet

9. What is the length of the common rafters to be ordered for section *B* if the roof slope is 6:12? *Note:* The overhang is 18".

    a. 9 ft.

    b. 10 ft.

    c. 11 ft.

    d. 12 ft.

10. Three-tab, 15# shingles are to be used for this residential building (3 bundles per square). The slope of the roof is 5:12, and the overhang is 1'. Waste should be based on 5%. How many bundles are needed?

    a. 72

    b. 81

    c. 91

    d. 97

PART TWO
Administration

Questions are based on the Building Code.

The following code quiz is based on topics that concern the building administration. Exam candidates preparing for an exam based on the *International Residential Code* and the *International Building Code* will find the answers in Chapter 1.

The code quizzes are written to familiarize you, the exam candidate, with the reference book. If you feel that you know the answer to the question, the correct answer is less important than your ability to identify the corresponding code location. Remember, these quizzes are representative of the subject matter covered on exams but do not represent actual questions. Do not attempt to memorize the answers. It is much more effective to understand how to locate the subject matter in the approved references.

Questions in this quiz cover the following subject matter:

- Permits
- Certificates of occupancy
- Demolition
- Temporary permits
- Building officials
- Construction documents

Notes

1. Which of the following statements is *true* regarding site plans submitted for demolition projects?

   a. Site plans are not required for structures to be removed.

   b. Site plans must show the construction to be demolished.

   c. Site plans must show the size and location of existing construction that will remain.

   d. Both b and c

2. A permit that has been issued is deemed void if work does not begin within ___ days of issuance.

   a. 30

   b. 60

   c. 90

   d. 180

3. In addition to one set of approved plans being retained by the building official, an approved copy must always be available:

   a. At the job site

   b. At the contractor's office

   c. When the inspector requests to see a copy

   d. At the county clerk's office

4. The building permit must be:

   a. Displayed in the contractor's office

   b. Posted on the job site

   c. Provided to the inspector when requested

   d. Posted in the building official's office

5. Which of the following construction projects would *not* require a building permit?

    a. A 400 square foot, two-story building

    b. A one story, 200 square foot home attached to an existing two-car garage

    c. A one-story detached accessory structure with 190 square foot

    d. None of the above

6. Who has the authority to approve the use of temporary power?

    a. Electrical contractor

    b. General contractor

    c. Home owner

    d. Building official

7. Which of the following is not found on a Certificate of Occupancy?

    a. If an automatic sprinkler is provided

    b. Name and address of the owner

    c. Name of the building official

    d. Contractor's license number

8. Which of the following *must* be included on a stop work order?

    a. Contractor's name and license number

    b. Name of person doing the work

    c. Conditions under which work will be permitted to resume

    d. Number of days the order will be in place

Notes

Notes

9. One set of approved construction documents must be retained by the building official for not less than how many days from the date of the completion?

    a. 90

    b. 180

    c. 30

    d. 120

10. One set of approved construction documents shall be retained for not less than 180 days by which of the following?

    a. Contractor

    b. Home owner

    c. Building official

    d. None of the above

PART THREE
Design and Planning

Questions are based on the Building Code.

The following code quiz is based on topics primarily effecting the design of a building. Exam candidates preparing for an exam based on the *International Residential Code* will find the answers in Chapter 3. Candidates using the *International Building Code* will find the answers in various chapters.

As stated previously, the code quizzes are written to familiarize you with the reference book. Do not attempt to rely on your field knowledge to pass the exam. It is imperative for you to become familiar with using the code book. These quizzes will prepare you for the exam in the most efficient manner. Take the time to locate the answer to each question using the code book.

Questions in this quiz cover the following subject matter:

- Ceiling heights
- Lighting and ventilation
- Means of egress
- Rescue opening
- Guardrails
- Stairs
- Sanitation
- Live Loads/Dead Loads

Notes

MIN 7'0" for:, Bath-rooms, toilet rooms, kitchens, storage rooms and laundry rooms

Sloped ceilings in half of the room only, any area less than 5' will not be computed as finished/heated square feet

MIN 7'6 Corridors,

MIN 5'

MIN 20" × 30" Attic Access

Every dwelling shall have at least one room with a MIN of 150 SF.

Each room, except the kitchen, shall be a MIN of 70 SF.

No habitable room, except the kitchen, shall measure less than 7' in any dimension.

Ceiling or wall access to attic must be provided with easily operable lid/door.

Access is required unless the clear height of the attic at the peak of the roof is less than 24".

IRC 304
IBC 1208

MIN HT:

24"

MIN Width: 20"

MIN net clear opening:

First floor net clear opening:

Sill height:

Min net clear opening = 5.7 sf
except grade level = 5.0

44"

IRC 310
IBC 1025

IBC 1009
IRC 311

MIN_____
Headroom must be
maintained

The total of consecutive riser,
tread, riser: 25" MAX
24" MIN

MAX

MIN

(exclusive of nosing)

3 1/2" MAX

4" Sphere should not fit through this area

6" Sphere should not fit through this area

Notes

1. For an emergency escape opening in a habitable space, the minimum net clear opening width is specified as _____ inches.

   a. 20

   b. 22

   c. 24

   d. 28

2. Open access from a garage into a room intended for sleeping purposes is prohibited and should be closed. If a solid wood door is installed, the minimum thickness of the door must be _____ inches.

   a. 1¼

   b. 1½

   c. 1⅜

   d. 2

3. The minimum uniformly distributed live load for a room with an intended use of sleeping is designated as _____ psf.

   a. 10

   b. 20

   c. 30

   d. 50

4. To test the allowable triangular-shaped opening created by the riser, tread, and bottom rail of a guard on an open stairway, a sphere of _____ inches in diameter cannot pass through.

   a. 4

   b. 6

   c. 8

   d. 10

5. For ramps intended for use other than specifically for means of egress, the maximum allowable slope is:

    a. 8%

    b. 10%

    c. 10.5%

    d. 12.5%

6. Untreated wood joists should not be installed closer than _____ inches.

    a. 12

    b. 14

    c. 16

    d. 18

7. Except for special types of stairways, such as spiral, a required headroom of _____ must be maintained in all parts of the stairway.

    a. 6 feet 6 inches

    b. 6 feet 8 inches

    c. 7 feet 0 inches

    d. 7 feet 6 inches

8. A smoke-developed index of no more than _____ should be maintained for all wall and ceiling finishes.

    a. 200

    b. 250

    c. 375

    d. 450

Notes

9. Except for kitchens, habitable rooms shall not be less than _____ feet in any horizontal dimension.

    a. 6

    b. 7

    c. 8

    d. 9

10. A 10' × 10' habitable room should have an aggregate glazed area of _____ square feet to provide natural lighting.

    a. 5

    b. 6

    c. 7

    d. 8

11. Not more than _____ of the required floor area of a room included as square feet is allowed to have a sloped ceiling height of 7'.

    a. 30%

    b. 40%

    c. 50%

    d. 60%

12. Foam plastic picture mold, chair rails, baseboards, handrails, ceiling beams, and similar trim are allowed by code provided the minimum density is:

    a. 15 pcf

    b. 20 pcf

    c. 25 pcf

    d. 30 pcf

13. All rescue openings are required to have at least a net clear opening of _____ square feet unless situated on a grade level.

    a. 5.7

    b. 5.0

    c. 4.7

    d. None of the above

14. The maximum flame spread is _____ for any structure's wall and ceiling finishes. This does not apply to most trim finishes.

    a. 75

    b. 150

    c. 200

    d. 250

15. In the event an emergency escape and rescue opening is positioned with a sill height below the adjacent grade, a window well is required. The minimum horizontal area of the well should be _____ square feet.

    a. 8

    b. 9

    c. 10

    d. 12

16. For the purpose of the code, a vehicle covering is considered a carport if it is open:

    a. On at least one side other than the entry

    b. On all four sides

    c. On at least two sides

    d. On at least one side and not adjacent to the structure

Notes

17. The minimum width of an exit passageway (hallway) for a single residence shall be at least:

    a. 3 feet

    b. 3 feet 6 inches

    c. 4 feet

    d. 4 feet 2 inches

18. A guardrail system is required for protection when a porch or balcony is located more than _____ inches above the grade or floor below.

    a. 24

    b. 30

    c. 36

    d. 40

19. What is the required width of a landing serving a 42"-wide stairway?

    a. 36"

    b. 42"

    c. 48"

    d. Any of the above

20. What size and type gypsum board is required to separate a garage from a habitable room located above it?

    a. ½" Type X

    b. ⅝" Type X

    c. ¾" Type X

    d. ³⁄₁₆" Type X

21. Additional safety measures, such as the use of tempered glass, are required when glazing adjacent to stairways is within _____ inches horizontally of the bottom tread of a stairway in any direction and the exposed surface of the glass is less than 5' above the nose of the tread.

    a. 36

    b. 42

    c. 48

    d. 60

22. Handrail height for stairs, when required, should be a minimum of _____ inches and a maximum of _____ inches measured vertically from the sloped plane adjoining the tread nosing.

    a. 38/34 (respectively)

    b. 34/36 (respectively)

    c. 34/38 (respectively)

    d. 36/38 (respectively)

23. Where entry is made to a crawl space for service of utilities, foam plastic insulation shall be protected against ignition by _____ inch-thick corrosion-resistant steel.

    a. 0.024

    b. 0.016

    c. 0.25

    d. 0.033

24. Wood siding installed on the exterior of a structure should have a clearance of _____ inches from the ground.

    a. 4

    b. 6

    c. 8

    d. 10

Notes

25. The minimum height of a parapet wall where the roof surface is adjacent to the wall or walls at the same elevation is:

   a. 24"

   b. 28"

   c. 30"

   d. 36"

PART FOUR
Foundations

Questions are based on the Building Code.

The following code quiz is based on topics and issues related to foundations. Exam candidates preparing for an exam based on the *International Residential Code* will find the answers in Chapter 4. Candidates using the *International Building Code* will find the answers in various chapters.

Questions in this quiz cover the following subject matter:

- Footings
- Foundation walls
- Drainage
- Waterproofing

Notes

1. A plain masonry hollow wall that is 7' tall with 6' of unbalanced backfill should be at least _____ thick. The soil type is GW.

   a. 6"

   b. 8"

   c. 10"

   d. 12"

2. The allowable foundation (load-bearing pressure) in pounds per square foot for sedimentary and foliated backfill is:

   a. 2,000

   b. 3,000

   c. 4,000

   d. 12,000

3. Where masonry veneer is used, concrete foundation walls are required to extend _____ inches above the finished grade adjacent to the foundation walls at all points.

   a. 2

   b. 4

   c. 6

   d. 8

4. A carport slab is to be installed in an area that is subjected to severe weather conditions. The minimum compressive strength of the concrete to be used is _____ psi.

   a. 2,000

   b. 2,500

   c. 3,000

   d. 3,500

5. All exterior footings should be placed at least _____ inches below the undisturbed ground surface.

   a. 6

   b. 8

   c. 10

   d. 12

6. Foundations plates or sills should be attached to concrete foundations with bolts:

   a. ½ inch in diameter with a penetration of at least 7 inches

   b. ¾ inch in diameter with a penetration of at least 7 inches

   c. ½ inch in diameter with a penetration of at least 6 inches

   d. ¾ inch in diameter with a penetration of at least 6 inches

7. The grade adjacent to a structure should be sloped away from the building to create a positive drain. The slope should not be less than:

   a. 5%

   b. 6%

   c. 8%

   d. 10%

8. Openings for under-floor (crawl space) ventilation should be covered provided that the least dimension of the covering does not exceed _____ inch(es).

   a. ¼

   b. ½

   c. ¾

   d. 1

Notes

Notes

9. Gravel or crushed stone drains installed for the purpose of foundation drainage shall extend at least _____ beyond the outside edge of the footing.

   a. 0.5 ft.

   b. 1 ft.

   c. 1.5 ft.

   d. 2 ft.

10. Concrete in footings should have a specified pounds per square inch of at least:

    a. 1,500

    b. 2,000

    c. 2,500

    d. 3,000

11. Stepped footings are required when the grade falls to cause the bottom surface of the footings to slope more than _____ unit(s) vertical and ten units horizontal.

    a. 0.5

    b. 1

    c. 1.5

    d. 2

12. All permanent supports are not required to be protected from frost. Which of the following is required to have frost protection?

    a. A freestanding building more than 400 square feet

    b. A freestanding building 400 square feet or less with an eave 10 feet in height

    c. A freestanding building 400 square feet or less with an eave 8 feet in height

    d. A freestanding building 300 square feet with an eave 8 feet in height

13. A wall 10" thick is being supported on a foundation. What is the minimum foundation required to support this wall?

    a. 8"

    b. 10"

    c. 12"

    d. 16"

14. Pier and curtain wall foundations are permitted to support light frame construction not more than two stories. All of the following requirements must be met by code except:

    a. Load-bearing wall shall be placed on continuous concrete footings placed integrally with the exterior wall

    b. Thickness of the load-bearing wall shall not be less than 3⅜" nominal

    c. Maximum height of a 4" load-bearing masonry foundation wall supporting wood framed walls shall be more than 4' in height

    d. Unbalanced fill for 4" foundation walls shall not exceed 24" for solid masonry.

15. A rubble stone foundation wall shall have a minimum thickness of _____ inches.

    a. 8

    b. 10

    c. 12

    d. 16

Notes

Notes

PART FIVE
Floors

Questions are based on the Building Code.

The following code quiz is based on topics and issues related to floor construction. Exam candidates preparing for an exam based on the *International Residential Code* will find the answers in Chapter 5. Candidates using the *International Building Code* will find the answers in various chapters.

Questions in this quiz cover the following subject matter:

- Joist bearing
- Joist sizing
- Boring and notching
- Spans
- Fastening
- Drainage
- Waterproofing

Notes

1. Joist framing from opposite sides of a beam or girder shall be lapped at least _____ inches.

    a. 2

    b. 3

    c. 4

    d. 6

2. A single header is allowed in a framed floor opening if the dimension along the header is less than _____ feet.

    a. 3

    b. 4

    c. 5

    d. 6

3. SPF #2 is to be installed 16" on center in a residential living area with a dead load of 20 psf. The length of the room is 14' 6". What size floor joist is required?

    a. 2 × 6

    b. 2 × 8

    c. 2 × 10

    d. 2 × 12

4. Holes bored in a floor joist to install plumbing pipes should not be placed closer than _____ from the top.

    a. Two times the thickness of the pipe

    b. 2 inches

    c. One half the joist depth

    d. None of the above

5. In residential structures, when there is usable space above and below the concealed space of a floor/ceiling assembly, draftstops must be installed so that the area of the concealed space does not exceed _____ square feet.

   a. 500

   b. 800

   c. 1,000

   d. 1,200

6. Unless bearing on a beam or similar support, approved hangers are required for framed openings in the floor when the header joist to trimmer joist connections and header joist span exceeds _____ feet.

   a. 4

   b. 5

   c. 6

   d. 12

7. The minimum thickness of a concrete slab placed on grade is _____ inches.

   a. 3

   b. 3.5

   c. 4

   d. 6

8. Sheathing, with a span rating of 24/0, to be used for subfloor can be installed with a maximum span of _____ inches.

   a. 0

   b. 16

   c. 24

   d. 32

Notes

Notes

9. What is the maximum span for a 2 × 8 Southern yellow pine #2 installed in the sleeping area of a structure? The dead load is 20 psf, and the joist spacing is 24 inches on center.

   a. 12 feet 4 inches

   b. 14 feet 2 inches

   c. 13 feet 6 inches

   d. 11 feet 0 inches

10. Beams and girders must bear at least _____ inch(es) on a masonry column.

    a. 1

    b. 1½

    c. 1¾

    d. 3

11. Joists framing into the side of a wood girder must be supported by:

    a. Framing anchors

    b. 2 × 2 Ledger strips

    c. All of the above

    d. None of the above

12. A notch made at the end of a floor joist should not exceed _____ the depth of the joist.

    a. ⅓

    b. ¼

    c. ⅙

    d. ½

13. A one-story building with a width of 28' requires _____ girder(s). *Note*: The span is 10' 2".

    a. Four 2 × 12

    b. Four 2 × 10

    c. Four 2 × 8

    d. Three 2 × 10

14. The required joint laps for approved vapor barrier to be placed between a concrete floor slab and the grade is _____ inches.

    a. 4

    b. 5

    c. 6

    d. 8

15. The minimum thickness of particle board underlayment shall be _____ inch(es).

    a. ¼

    b. ½

    c. ¾

    d. 1

16. What is the maximum span for a 2 × 8 nominal floor joist (Southern pine #2) to be installed in a sleeping area? *Note*: Dead load is 10 psf, and joists will be installed 16 inches on center.

    a. 10 feet 9 inches

    b. 14 feet 2 inches

    c. 13 feet 6 inches

    d. 13 feet 3 inches

Notes

Notes

17. What is the maximum notch depth allowed for a 2 × 12 nominal floor joist? *Note:* The joist spans 14 feet 6 inches, and the notch needs to be made 5 feet from one end.

    a. 1⅞ inches

    b. 1¾ inches

    c. 2 inches

    d. None of the above

18. Southern yellow pine #2 is to be installed in a residential sleeping area 16" on center. The span will be 18', and the dead load is 10 psf. What size floor joist is required?

    a. 2 × 6

    b. 2 × 8

    c. 2 × 10

    d. 2 × 12

19. What type lumber is required for a sleeping area with a 20 psf dead load if a 2 × 10 is to span 18' with 16" on center spacing? *Note:* The live load is 30 psf.

    a. Douglas fir-larch ss

    b. Hem-fir ss

    c. Southern pine ss

    d. None of the above

20. A hole to be bored into a floor joist for the purpose of running a plumbing pipe or electrical conduit:

    a. Must not exceed one third the depth of the joist

    b. Must not exceed one sixth the depth of the joist

    c. Cannot be bored in the middle third of the joist

    d. Is not allowed unless approved by a qualified engineer

21. How many jack studs are required to support three 2 × 12s that span 10' 2" on a 20' wide building?

    a. One on each end

    b. Two on each end

    c. Three on each end

    d. Four on each end

22. When joists exceed a nominal 2 × 12" (6:1 depth-to-thickness ratio), bridging must be installed for each _____ feet of span.

    a. 4

    b. 8

    c. 12

    d. 16

23. Without approval from a registered design professional, which of the following jobsite modifications can be made to wood floor trusses?

    a. Cut

    b. Spliced

    c. Notched

    d. None of the above

24. Vapor retarder is not required between the base course / subgrade and a concrete floor slab in which of the following circumstances?

    a. Typical driveway and walkways

    b. Unheated accessory structures

    c. For installations approved by the building official based on local site conditions

    d. All of the above

Notes

25. The specified residential sleeping area live load is _____ psf.

    a. 10

    b. 20

    c. 30

    d. 40

PART SIX

Walls

Questions are based on the Building Code.

The following code quiz is based on topics and issues related to wall construction. Exam candidates preparing for an exam based on the *International Residential Code* will find the answers in Chapter 6. Candidates using the *International Building Code* will find the answers in various chapters.

Questions in this quiz cover the following subject matter:

- Stud spacing
- Cutting and notching
- Bearing and support
- Fastening

Notes

1. Where rafters are spaced more than 16 inches on center and the bearing studs below are spaced 24 inches on center, members shall bear within _____ inches of the studs beneath.

   a. 4

   b. 5

   c. 6

   d. 7

2. What is the maximum spacing of a 2 × 6 for a bearing wall to support a roof and ceiling only?

   a. 16 inches on center

   b. 20 inches on center

   c. 24 inches on center

   d. 28 inches on center

3. Where horizontal steel strap blocking is used to provide edge blocking, it shall be at least _____ inch(es) wide.

   a. 1

   b. 1¼

   c. 1½

   d. 2

4. What is the maximum stud spacing if particleboard wall sheathing is to be nailed to the stud walls? *Note*: M-2 Exterior glue grade, ½ inch thickness.

   a. 12 inches on center

   b. 16 inches on center

   c. 24 inches on center

   d. 30 inches on center

5. Fireblocking is required horizontally at intervals not exceeding _____ feet to cut off all concealed draft openings in framed walls in order to form an effective fire barrier between stories.

   a. 5

   b. 8

   c. 10

   d. 12

6. Foundation cripple walls are required to be framed of studs not less in size than the studding above with a minimum length of _____ inches.

   a. 10

   b. 12

   c. 14

   d. 16

7. When notching of a top plate is required for the purpose of placing piping or ductwork, a galvanized metal tie must be fastened to each plate across and to each side of the opening. The metal tie must be fastened with _____ nails and must be _____ wide.

   a. 12d/½ inch

   b. 12d/1½ inch

   c. 16d/½ inch

   d. 16d/1½ inch

8. A hole may not be bored closer than _____ inch to the edge of a stud located in an exterior load-bearing framed wall.

   a. ¼

   b. ½

   c. ¾

   d. ⅝

9. The minimum thickness of cold-formed steel studs is _____ mils.

   a. 0.024

   b. 0.033

   c. 0.043

   d. 0.054

10. Fasteners for attaching structural sheathing (shear panels) to steel wall framing must be installed with a minimum edge distance of _____ inch.

    a. ¼

    b. ½

    c. ⅜

    d. 1

11. The minimum flange width for a load-bearing cold-formed steel stud is _____ inches.

    a. 1½

    b. 1⅝

    c. 1¾

    d. 2

12. Special instructions apply for splicing and cripple wall installations where stepped footings cause the height of a required braced wall panel extending from the foundation to the floor above to vary more than _____ feet.

    a. 4

    b. 6

    c. 8

    d. 10

13. Buildings must be provided with exterior and interior braced wall lines. The spacing of the braced wall lines shall not exceed _____ feet in both the longitudinal and transverse directions in each story. *Note*: No exceptions apply.

    a. 10

    b. 25

    c. 35

    d. 50

14. The maximum-size hole that can be placed in any stud is _____ inches.

    a. 1.4

    b. 1.6

    c. 2

    d. 2.2

15. What is the maximum spacing of 2 × 4's to be installed in a bearing wall to support one floor, roof, and ceiling?

    a. 10"

    b. 14"

    c. 16"

    d. 24"

16. Particleboard (M-1/M-S Grade) wall sheathing nailed to 16 inches on center studs must be at least _____ inch(es) in thickness.

    a. ½

    b. ⅜

    c. ⅝

    d. 1

Notes

17. A 4-foot opening to be located in an interior nonbearing partition may be constructed with a _____ header. *Note:* Vertical distance above the opening to the parallel nailing surface above is 24 inches.

    a. Single stud on each side to support a flat 2 × 4

    b. Double stud, single header

    c. Double stud supporting two flat 2 × 4's

    d. None of the above

18. Interior partitions (nonbearing) may be constructed with at least a:

    a. Double top plate

    b. Single top plate

    c. Double top plate if wall height exceeds 10 feet

    d. None of the above

19. Cripple walls having a stud height exceeding _____ inches shall be braced or shall be framed of solid blocking.

    a. 14

    b. 24

    c. 36

    d. 48

20. When single top plates are permitted, the plate must be adequately tied at joints, corners, and intersecting walls by a galvanized steel plate that is nailed to each wall or segment by _____ nails on each side.

    a. Eight 16d

    b. Four 16d

    c. Six 8d

    d. Four 8d

21. The minimum-size screw to attach steel sheets to metal framing (steel to steel) is:

    a. #5

    b. #6

    c. #7

    d. #8

22. What is required when nailing a top plate to a stud?

    a. Two 16d common

    b. Three 16d common

    c. Two 8d common

    d. Three 8d common

23. Which of the following materials is *not* allowed for fireblocking required in wall framing?

    a. 2" nominal lumber

    b. ½" particleboard with joints backed by ½" particleboard

    c. Properly installed ½" gypsum board

    d. Two thicknesses of 1" nominal lumber with broken lap joints

24. Any stud in an exterior wall or bearing partition may be cut or notched to a depth not to exceed _____ of its width. *Note*: No exceptions.

    a. 15%

    b. 25%

    c. 40%

    d. 60%

Notes

25. Utility-grade studs cannot be spaced more than _____ inches on center.

    a. 12

    b. 16

    c. 20

    d. 24

PART SEVEN

Wall Coverings

Questions are based on the Building Code.

The following code quiz is based on topics and issues related to wall covering. There are two quizzes in this section, one based on the IRC and one based on the IBC. Exam candidates preparing for an exam based on the *International Residential Code* will find the answers in Chapter 7. Candidates using the *International Building Code* will find the answers in Chapter 14.

Questions in this quiz cover the following subject matter:

- Gypsum
- Masonry veneer
- Aluminum siding
- Plaster

Notes

The following quiz is based on the *International Residential Code.*

1. What is the maximum support spacing for ½" gypsum on walls or ceilings?

 a. 16"

 b. 24"

 c. 30"

 d. 36"

2. Shakes used for exterior covering should be held in place by which of the following type fasteners?

 a. Stainless steel

 b. Aluminum nails

 c. Staples

 d. All of the above

3. What is the nominal thickness of horizontal aluminum siding with insulation?

 a. 0.019 inch

 b. 0.024 inch

 c. $7/16$ inch

 d. 2 inches

4. A weep screed should be provided at or below the foundation plate line on exterior stud walls and shall be placed a minimum of _____ inch(es) above paved areas.

 a. 1

 b. 2

 c. 3

 d. 4

5. What is the minimum number of plaster coats required when applied over metal lathe?

    a. 1

    b. 2

    c. 3

    d. 4

6. Approved corrosive-resistive flashing should be provided in the exterior wall envelope at all the following locations *except*:

    a. At wall and roof intersections

    b. At built-in gutters

    c. Over metal copings and sills

    d. Continuously above all projecting wood trim

7. Gypsum board utilized as the base or backer for adhesive application of ceramic tile should conform to:

    a. ASTM C 1178

    b. ASTM C 94

    c. ASTM C 1395

    d. ASTM C 1047

8. What is the maximum spacing of screws allowed for ⅜" gypsum board installed on a wall with adhesive?

    a. 7"

    b. 12"

    c. 16"

    d. 24"

Notes

9. Water-resistant gypsum board should *not* be used in which of the following locations?

   a. Shower

   b. Bathtub

   c. Area subject to high humidity

   d. All of the above

10. Masonry veneer lintels should *not* support any vertical load other than:

   a. Live loads

   b. Dead loads

   c. Nonloadbearing

   d. Loadbearing

The following quiz is based on the *International Building Code.*

1. Stone veneer units not exceeding 10" in thickness should be anchored directly to each of the following *except*:

   a. Masonry

   b. Concrete

   c. Stud construction

   d. All of the above

2. Exterior walls should provide weather protection for the building. Which of the following types of covering is required to have a minimum thickness of ¼"?

   a. Anchored masonry veneer

   b. Vinyl siding

   c. Adhered terra cotta

   d. Precast stone facing

3. The height of any section of thin exterior structural glass veneer should not exceed _____ inches.

   a. 12

   b. 18

   c. 36

   d. 48

4. What is the maximum weight allowed by code of interior adhered masonry veneers?

   a. 15 psf

   b. 20 psf

   c. 25 psf

   d. 30 psf

Notes

5. A mixture of calcined gypsum, or calcined gypsum, lime, and aggregate is defined as:

   a. Gypsum plaster

   b. Cement plaster

   c. Gypsum veneer plaster

   d. None of the above

6. Wood veneers on exterior walls should not project more than _____ inches from the building wall.

   a. 12

   b. 18

   c. 24

   d. 30

7. Nails used to fasten vinyl siding and accessories should have a minimum shank diameter of _____ inch(es).

   a. 0.313

   b. 0.125

   c. 0.75

   d. 1

8. Vinyl siding installed vertically should have fastener spacing not to exceed _____ inches.

   a. 12

   b. 16

   c. 18

   d. 24

9. Flashing should be installed in which of the following locations?

    a. Chimneys

    b. Decks

    c. Built-in gutters

    d. All of the above

10. Fiber cement horizontal lap siding should be lapped a minimum of
    _____ inch(es) and located over a strip of flashing.

    a. ½

    b. ¾

    c. 1

    d. 1¼

Notes

Notes

PART EIGHT
Roof Framing

Questions are based on the Building Code.

The following code quiz is based on topics and issues related to roof framing construction. Exam candidates preparing for an exam based on the *International Residential Code* will find the answers in Chapter 8. Candidates using the *International Building Code* will find the answers in Chapter 23.

Questions in this quiz cover the following subject matter:

- Rafter and joist spacing
- Fastening
- Cutting and notching
- Ventilation
- Access

Notes

1. For roof framing, rafters, and ceiling joists, when the nominal depth-to-thickness ratio of the framing member exceeds 6:1, bridging must be installed every _____ feet.

    a. 6

    b. 7

    c. 8

    d. 10

2. A 2 × 6 SYP #2 material is to be used for the ceiling joists with a live load of 10 psf. The joists, spaced 24 inches on center, will be installed to support an attic that will not be used for storage. What is the maximum span?

    a. 11 feet 0 inches

    b. 15 feet 6 inches

    c. 15 feet 11 inches

    d. 17 feet 8 inches

3. Ceiling joists shall have a bearing surface of not less than _____ inch(es) on the wood top plate.

    a. 1

    b. 1½

    c. 2

    d. 2½

4. For ceilings not attached to rafters with a roof live load of 20 psf and a dead load of 10 psf, the maximum span allowed for a 2 × 8 rafter to be spaced 16 inches on center is _____. *Note:* Material is SYP #2.

    a. 18 feet 6 inches

    b. 19 feet 5 inches

    c. 14 feet 4 inches

    d. 19 feet 3 inches

5. For ceilings attached to rafters with a roof live load of 20 psf and a dead load of 20 psf, the maximum span allowed for a 2 × 8 rafter to be spaced 12 inches on center is _____. *Note:* Material is Douglas Fir #1.

   a. 18 feet 6 inches

   b. 19 feet 5 inches

   c. 14 feet 4 inches

   d. 19 feet 3 inches

6. A notch must be placed in a ceiling joist. The nominal size of the joist is 2 × 10 and spans 12 feet 8 inches. What is the maximum notch that can be placed on the top, 2 feet from the exterior wall?

   a. 1.67 inches

   b. 1.54 inches

   c. 1.33 inches

   d. 0 inches

7. How many 16d common nails are required for a rafter connection if the roof slope is 4:12? Rafters are 16" on center, the roof span is 20', and the ground snow load is 30 psf.

   a. 4

   b. 6

   c. 8

   d. 10

8. What type of nails can be substituted for 16d nails when fastening rafter connections?

   a. 10d

   b. 12d

   c. 40d box

   d. No substitutions provided by code

Notes

9. A 24' rafter spaced 24" on center must have a connection strength for resisting uplift winds of _____ pounds considering a basic wind speed of 90 mph.

   a. –145

   b. –181

   c. –262

   d. –351

10. When a ceiling joist must be spliced, what is the least amount it must be lapped?

    a. 2"

    b. 3"

    c. 4"

    d. 6"

11. Purlins used to support roof loads must be at least:

    a. 2 × 6

    b. One size larger than the rafter it is bracing

    c. No less than one size smaller than the rafter it is bracing

    d. The size of the rafter it is supporting

12. For buildings with a roof pitch of less than _____, hips and valleys must be designed as beams.

    a. 2:12

    b. 3:12

    c. 4:12

    d. 5:12

13. A fireplace opening that requires a header greater than _____ must be supported by approved rafter hangers.

    a. 4'

    b. 5'

    c. 6'

    d. 7'

14. What is the minimum net thickness of roof sheathing (surfaced dry) if the framing is 24" on center?

    a. ½"

    b. ⅜"

    c. ⅝"

    d. ⁷⁄₁₆"

15. A hole must be bored in a ceiling joist and needs to be placed near the edge. What is the closest the hole can be placed to the edge?

    a. 1"

    b. 2"

    c. 2½"

    d. 3"

16. The unbraced length of braces supporting purlins shall not exceed _____.

    a. 4'

    b. 6'

    c. 8'

    d. 10'

Notes

Notes

17. What species of lumber must be used if 2 × 6 nominal ceiling joists are to span 14' 8" and be spaced 16" on center? The live load is 20 psf, and the application is uninhabitable attics with limited storage.

    a. Douglas fir-larch

    b. Hem-fir

    c. Southern pine

    d. Spruce-pine-fir

18. What is the maximum span for 2 × 8 ceiling joists, 16 inches on center, to support an uninhabited attic with limited storage? *Note:* Dead load is 10 psf, and roof live load is 20 psf.

    a. 23 feet 6 inches

    b. 21 feet 7 inches

    c. 20 feet 10 inches

    d. 19 feet 7 inches

19. What is the minimum required headroom (unobstructed) in the attic space above the access opening?

    a. 30"

    b. 36"

    c. 48"

    d. 72"

20. Each of the following must be listed on the label for fire-retardant-wood *except:*

    a. ID mark of the approving agency and treating manufacturer

    b. Name of the fire-retardant treatment and species of wood

    c. Flame and smoke spread index

    d. Statement of compliance with ASTM E 184

21. The net free ventilation area for attics must be at least _____ of the area. *Note:* No exceptions.

    a. ¹⁄₁₅₀

    b. ¹⁄₂₀₀

    c. ¹⁄₂₅₀

    d. ¹⁄₃₀₀

22. The maximum-allowed moisture content for lumber used to frame a 4:12 pitch roof is:

    a. 15%

    b. 19%

    c. 22%

    d. 27%

23. A listed flame spread index of _____ or less is required when codes specify fire-retardant-treated wood.

    a. 15

    b. 20

    c. 25

    d. 30

24. To prevent ventilation obstruction, _____ inch(es) of space must be maintained between the insulation and sheathing where eave or cornice vents are installed.

    a. 1

    b. 1½

    c. 2

    d. 2¾

Notes

Notes

25. What is the maximum span of a 2 × 6 SPF #2 ceiling joist to be spaced 12 inches on center? The attic will not be utilized in any way. Dead load is 5 psf, and the live load is 10 psf.

    a. 18 feet 8 inches

    b. 14 feet 9 inches

    c. 16 feet 3 inches

    d. 13 feet 9 inches

PART NINE

Roof Coverings

Questions are based on the Building Code.

The following code quiz is based on topics and issues related to roofing assemblies. Exam candidates preparing for an exam based on the *International Residential Code* will find the answers in Chapter 9. Candidates using the *International Building Code* will find the answers in Chapter 15.

Questions in this quiz cover the following subject matter:

- Asphalt shingles
- Wood shakes
- Mineral-surfaced roll roofing
- Underlayment
- Attachment
- Valley
- Flashing

Notes

1. What is the minimum slope of asphalt shingles that may be used on a roof?

   a. 2:12

   b. 3:12

   c. 4:12

   d. 6:12

2. All of the following fasteners are used for asphalt shingles *except:*

   a. Galvanized steel

   b. Copper

   c. Aluminum

   d. Nickel

3. The minimum-diameter head permitted to be on a fastener for asphalt shingles is _____ inches.

   a. 0.105

   b. 0.375

   c. 0.50

   d. 0.75

4. Asphalt shingles for normal application shall be secured with not less than _____ fastener(s) per shingle.

   a. 1

   b. 2

   c. 3

   d. 4

5. *Given:* A roof slopes 17%, and underlayment is being installed. The first layer of underlayment is a minimum of 19 inches wide, and the second layer shall be _____ inches wide.

   a. 14

   b. 24

   c. 36

   d. 42

6. Fasteners for underlayment installed in an area with high winds shall be applied along the overlap not farther than _____ inches apart on center.

   a. 12

   b. 24

   c. 36

   d. 42

7. On roofs sloped 4:12 or greater, underlayment shall be at least one-layer applied shingle fashion starting from the eaves and lapped _____ inch(es).

   a. 1

   b. 2

   c. 4

   d. 5

8. Base flashing shall have a minimum thickness of _____ inch.

   a. 0.016

   b. 0.017

   c. 0.018

   d. 0.019

9. For closed valleys, valley lining of one ply of smooth roll roofing must be at least _____ inches wide.

   a. 18

   b. 24

   c. 36

   d. 42

10. Clay tile shall be installed on a roof with a minimum slope of:

    a. 2½:12

    b. 3:12

    c. 4:12

    d. 5:12

11. Attachment wire for clay or concrete tile shall not be smaller than _____ inch.

    a. 0.062

    b. 0.077

    c. 0.083

    d. 0.094

12. Which of the following is a consideration when applying tile?

    a. Climate conditions

    b. Roof slope

    c. Type of tile

    d. All of the above

Notes

13. When installing metal roof shingles, an ice barrier is required if the average temperature in January is _____ degrees F.

    a. 25

    b. 30

    c. 32

    d. 35

14. When using metal roof shingles, the minimum height of a splash diverter required for roof valley flashing is _____ inch.

    a. ¼

    b. ½

    c. ¾

    d. 1

15. When using spaced sheathing for wood shingles, what are the minimum dimensions required?

    a. 1" × 4"

    b. 1½" × 3"

    c. 1" × 5"

    d. 1½" × 5"

16. What is the minimum head lap allowed for slate shingles secured to a roof with 6:12 slope?

    a. 1"

    b. 2"

    c. 3"

    d. 4"

Notes

17. Wood shakes shall be laid with a minimum side lap not less than _____ inch(es).

    a. ¾

    b. 1

    c. 1¼

    d. 1½

18. Which of the following types of wood shakes are not approved for grade No. 1?

    a. 18" shakes of naturally durable wood with 7.5" exposure

    b. 24" taper-sawn shakes of naturally durable wood with 7.5" exposure

    c. 24" taper-sawn shakes of naturally durable wood with 10" exposure

    d. 18" preservative-treated taper shakes of Southern yellow pine with 7.5" exposure

19. Which of the following types of fasteners shall be used for copper roofs?

    a. Galvanized

    b. Copper

    c. Stainless steel

    d. 300 Series stainless steel

20. A metal roof of hard lead shall have a standard thickness of:

    a. 4 psf

    b. 3 psf

    c. 2 psf

    d. 1 psf

21. Mineral-surfaced roll roofing shall not be applied on roof slopes with less than _____ slope.

    a. 2%

    b. 8%

    c. 12%

    d. 33%

22. What is the minimum design slope of a coal-tar built-up roof for drainage?

    a. 8%

    b. 6%

    c. 2%

    d. 1%

23. A sprayed polyurethane foam roof requires a protective coating to be applied no later than _____ after foam application.

    a. 2 hours

    b. 12 hours

    c. 72 hours

    d. Can be applied immediately

24. New roof coverings shall not be installed without first removing all existing layers of roof covering *except:*

    a. Where roof covering is water soaked

    b. Where clay tile is installed over existing wood shakes

    c. Where existing roof has two or more applications of roof covering

    d. Where existing roof covering is asbestos-cement tile

Notes

25. Coal-tar saturated organic felt used for built-up roof cover must conform to what standard?

   a. ASTM D 227

   b. ASTM D 450

   c. ASTM D 43

   d. ASTM D 4022

PART TEN
Chimneys and Fireplaces

Questions are based on the Building Code.

The following code quiz is based on topics and issues related to chimney and fireplace construction. Exam candidates preparing for an exam based on the *International Residential Code* will find the answers in Chapter 10. Candidates using the *International Building Code* will find the answers in Chapter 21.

Questions in this quiz cover the following subject matter:

- Chimney termination
- Flue lining
- Hearth size
- Footings
- Smoke chambers
- Firebox
- Framing

1. Footings for masonry chimneys in areas not subjected to freezing shall be at least _____ inches below finished grade.

   a. 6

   b. 12

   c. 18

   d. 24

2. A chimney wall may be reduced in size 4" above where the chimney passes through the roof if:

   a. Proper measurements are taken to prevent excess air from entering

   b. All combustible materials are more than 12" away from chimney wall

   c. The chimney wall is lined with a fire-retardant material

   d. Not permitted under any circumstance

3. Masonry chimneys constructed of hollow masonry units shall be grouted solid with not less than _____ inches of nominal thickness.

   a. 2

   b. 3

   c. 4

   d. 6

4. Masonry chimneys shall be lined appropriately for the type of appliance connected. Residential-type appliances having a clay flue lining must comply with the requirements of:

   a. ASTM C 27

   b. ASTM C 199

   c. ASTM C 315

   d. ASTM 1261

5. Where a spark arrestor is installed on a masonry chimney, the arrestor screen shall have heat and corrosion resistance equivalent to:

   a. 19-gauge stainless steel

   b. 24-gauge stainless steel

   c. 28-gauge stainless steel

   d. 30-gauge stainless steel

Notes

6. Chimney terminations shall be not less than _____ feet above the highest point where the chimney passes through the roof.

   a. 2

   b. 3

   c. 6

   d. 10

7. When installing flue lining, the lining shall be carried up vertically with a maximum slope no greater than _____ degrees from the vertical.

   a. 30

   b. 35

   c. 40

   d. 45

8. A 10" round chimney flue venting multiple appliances shall have a cross-sectional area of _____ square feet.

   a. 28

   b. 38

   c. 50

   d. 78

9. The upper edge of a chimney cleanout shall be located at least ____ inches below the lowest chimney inlet opening.

   a. 2

   b. 4

   c. 6

   d. 8

10. Chimneys that pass through the soffit or cornice shall have a minimum air space clearance of _____ inch(es).

    a. 1

    b. 2

    c. 3

    d. 4

11. *Given:* The floor of combustion chamber to the top of the flue is 18 feet. The fireplace opening is 2,200 square inches. What is the minimum cross-sectional area required for a retangular chimney flue?

    a. 125 square inches

    b. 140 square inches

    c. 168 square inches

    d. 214 square inches

12. Fireblocking shall be to a depth of 1" and shall be placed on strips of metal between chimneys and:

    a. Wood joists

    b. Beams

    c. Headers

    d. All of the above

13. Footings for masonry fireplaces shall extend at least ____ inches beyond the face of the fireplace.

    a. 4

    b. 6

    c. 10

    d. 12

14. Each strap used for seismic anchorage shall be fastened to a minimum of four floor-ceiling or floor joists using:

    a. one 1½" bolt

    b. two 1½" bolts

    c. one ½" bolt

    d. Two ½" bolts

15. What is the minimum depth of a masonry chimney firebox?

    a. 20"

    b. 24"

    c. 36"

    d. 30"

16. A fireplace throat shall be located a minimum of _____ inches above the lintel.

    a. 4

    b. 6

    c. 8

    d. 10

Notes

17. Smoke chamber walls shall be constructed of which of the following?

    a. Grouted stone

    b. Grouted concrete

    c. Hollow masonry units

    d. All of the above

18. The inside of a fireplace opening is 4 square feet. The inside height of the smoke chamber from the fireplace throat to the beginning of the flue shall not be greater than _____ square feet.

    a. 3

    b. 4

    c. 5

    d. 6

19. What is the absolute minimum thickness of a hearth extension?

    a. ⅜"

    b. 2"

    c. 4"

    d. 8"

20. A hearth extension shall extend at least 20 inches in front of the fireplace opening if the opening is:

    a. Larger than 2 sq. ft.

    b. Larger that 4 sq. ft.

    c. Less than 6 sq. ft.

    d. At least 6 sq. ft.

21. Exterior air intake shall not be located in any of the following areas *except*:

    a. A garage

    b. Outside a dwelling

    c. A basement

    d. An attic

22. Many wythes must be built between adjacent flue linings and shall be at least ___ inches thick and bonded into the walls of the chimney.

    a. 3

    b. 4

    c. 5

    d. 6

23. What is the cross section of a rectangular flue size with a 7½ inches × 15½ inches outside dimension?

    a. 58 square inches

    b. 74 square inches

    c. 82 square inches

    d. 91 square inches

24. Chimneys located inside a building shall have a minimum airspace combustibles of _____ inches.

    a. 2

    b. 4

    c. 6

    d. 12

25. A chimney shall not be corbelled more than _____ times the chimney's wall thickness from a wall.

    a. 0.25

    b. 0.375

    c. 0.5

    d. 0.75

PART ELEVEN

Concrete

Questions are based on general knowledge of concrete.

The following quiz is based on topics and issues based on concrete theory. As opposed to questions based on code, these address the why and how. Refer to your candidate information bulletin for the approved reference(s) that relate to concrete. Each of the questions in this quiz is based on general information about concrete, which can be found in many approved reference books. The answers to these questions will *not* be answered in the code book.

Questions in this quiz cover the following subject matter:

- Types of cement
- Joints
- Curing
- Forms
- Reinforcement
- Slump
- Cover

Notes

1. Typically, the required cover for reinforcement in concrete footings is _____ inches.

   a. 1½

   b. 2

   c. 2½

   d. 3

2. The space between the end of the rebar and the form is referred to as _____ .

   a. cover

   b. bond

   c. space

   d. void

3. For each inch of reduced slump in concrete, the pounds per square inch is reduced by _____ .

   a. 100

   b. 125

   c. 150

   d. 200

4. The load limit at which steel breaks is _____ .

   a. Yield strength

   b. Ultimate tensile strength

   c. Bond strength

   d. None of the above

5. The best type of Portland cement to use when the strength gain of concrete should be rapid is _____ .

   a. Type I

   b. Type II

   c. Type III

   d. Type IV

6. Concrete is made up of how many parts?

   a. 1

   b. 2

   c. 3

   d. 4

7. What is the diameter of a #3 deformed piece of steel reinforcement?

   a. 3"

   b. ⅓"

   c. ⅜"

   d. 0.33"

8. How many inches is specified for the splice if plans indicate that a #4 rebar is to be lapped 30D?

   a. 15"

   b. 4"

   c. 30"

   d. 18"

Notes

Notes

9. What type of nail is recommended for the installation of form material that will be removed?

   a. 8d Common

   b. 16d Common

   c. Double-headed

   d. Any of the above

10. The most common type of reinforcement used for the installation of footings is _____ .

   a. Deformed steel

   b. Smooth steel

   c. Rigid steel

   d. #3 Smooth steel

11. What admixture is used to achieve a high slump without the reduced pounds per square inch?

   a. Retarder

   b. Calcium

   c. Superplasticizer

   d. All of the above

12. Normal concrete is designed to reach its intended compressive strength within ____ days.

   a. 7

   b. 14

   c. 21

   d. 28

13. The type of concrete joint that will allow vertical as well as horizontal movement and would likely be used around the perimeter of a concrete column is:

    a. Construction

    b. Isolation

    c. Control

    d. Expansion

14. Typical rebar comes in _____ foot lengths that can be cut or bent on the job site according to needs specified.

    a. 10

    b. 15

    c. 20

    d. 25

15. Which of the following is *not* considered an admixture?

    a. Calcium

    b. Retarder

    c. Aggregate

    d. Superplasticizer

16. The most common welded-wire fabric used for residential applications has wires spaced _____ inches apart in each direction.

    a. 2

    b. 4

    c. 6

    d. 8

Notes

17. What type of form extrudes concrete in the shape of the form as it is pulled vertically or horizontally?

    a. Jump

    b. Slip

    c. Column

    d. Panelized

18. When pile foundations are cost-prohibitive due to the allowable bearing capacity being low to great depths, which of the following types of foundations are recommended?

    a. Floating

    b. Raft

    c. Mat

    d. Any of the above

19. In extremely hot weather, what is the maximum allowable time usually specified for the delivery time of ready-mixed concrete?

    a. 30 minutes

    b. 60 minutes

    c. 90 minutes

    d. 120 minutes

20. What type of cement is most commonly used for mixing concrete and is intended as general purpose?

    a. Type I

    b. Type II

    c. Type III

    d. Type IV

21. It is estimated the concrete formwork can account for as much as ___ of the cost for a concrete structure.

    a. 20%

    b. 40%

    c. 60%

    d. 75%

22. A drop chute must be used for concrete that will be dropped more than ___ feet.

    a. 3

    b. 4

    c. 5

    d. 6

23. Longitudinal joints installed to control cracking caused by temperature variations in pavement slabs are called _____ joints.

    a. warping

    b. construction

    c. isolation

    d. expansion

24. To increase the slump of concrete by 1 inch, the rule of thumb is to add ___ gallon(s) of water.

    a. 1

    b. 1¼

    c. 1½

    d. 1¾

Notes

25. Which of the following admixtures is added to concrete in hot weather for the purpose of providing more set time for placement?

    a. Calcium

    b. Retarder

    c. Water reducer

    d. Accelerator

PART TWELVE
Masonry

Questions are based on general knowledge of masonry.

The following quiz is based on topics and issues based on masonry theory. As opposed to questions based on code, these address the why and how. Refer to your candidate information bulletin for the approved reference(s) that relate to masonry. The answers to these questions will *not* be answered in the code book.

Questions in this quiz cover the following subject matter:

- Types of mortar
- CMU makeup
- Brick application
- Nominal size Versus Actual size
- Types of joints
- Bonds

1. The type of mortar best suited for use below grade is:

   a. Type K

   b. Type M

   c. Type N

   d. Type S

2. The type of joint that does not require jointing is:

   a. Flush

   b. Extruded

   c. Rough cut

   d. Weathered

3. The process of remixing mortar after it has started to become stiff and unusable is called:

   a. Reworked

   b. Dehydrated

   c. Rewet

   d. Retempered

4. With regard to masonry units, a "nominal" dimension refers to:

   a. Actual dimension

   b. Standard brick sizes

   c. Nominal brick plus the mortar joint

   d. Actual brick size plus the mortar joint

5. The process of placing a masonry unit beyond the vertical placement of the unit below is called:

   a. Benching

   b. Stepping

   c. Corbelling

   d. None of the above

6. Unused mortar should be discarded _____ hour(s) after it has been mixed.

    a. 1

    b. 1.5

    c. 2

    d. 2.5

7. Which of the following types of brick bond patterns would be considered the weakest?

    a. Flemish

    b. Running

    c. Stack

    d. English

8. A white powderlike deposit that develops on masonry walls that is caused by water-soluble salts is called:

    a. Efflorescence

    b. Lime build-up

    c. Mortar residue

    d. None of the above

9. What percentage of block, by weight, does aggregate make up for concrete masonry units?

    a. 50%

    b. 75%

    c. 80%

    d. 90%

Notes

10. What is the nominal dimension of a standard concrete masonry unit (block)?

    a. $7\frac{5}{8}$" × $7\frac{5}{8}$" × 16"

    b. $7\frac{5}{8}$" × 8" × $15\frac{5}{8}$"

    c. 8" × $7\frac{5}{8}$" × $15\frac{5}{8}$"

    d. 8" × 8" × 16"

11. Which of the following is the correct name for a block designed with molded or sawed openings for the purpose of absorbing sound?

    a. Insulating block

    b. Sound block

    c. Screen block

    d. Decorative block

12. A _____ masonry wall is constructed with two wythes that can react independently of one another and are separated by a continuous airspace of at least 2 inches (typical).

    a. veneer

    b. hollow

    c. cavity

    d. composite

13. What type of mortar is only suitable for use in nonload-bearing masonry?

    a. Type O

    b. Type N

    c. Type S

    d. Type M

14. Hydrated lime that has been formed into putty before packaging is called:

    a. Mixed

    b. Prehydrated

    c. Slaked

    d. Preserved

15. The recommended minimum time for machine-mixing ingredients for mortar is _____ minutes.

    a. 1 to 3

    b. 3 to 5

    c. 5 to 7

    d. 7 to 9

16. The typical mortar joint size to lay concrete bricks is ____ inch(es).

    a. ½

    b. ⅜

    c. ¾

    d. 1

17. What grade of concrete brick is required when the structure will be exposed to severe frost?

    a. Grade N

    b. Grade S

    c. Grade M

    d. Grade I

Notes

18. Which of the following is used to describe masonry bond?

    a. Mortar joint

    b. Pattern bond

    c. Structural bond

    d. All of the above

19. What type of mortar joint is best when the structure will be subject to heavy rains and high winds?

    a. Weathered

    b. Struck

    c. Flush

    d. V-shaped

20. A brick cut lengthwise across the end is called a _____ closure.

    a. queen

    b. king

    c. bat or half

    d. split

21. Which of the following grades of brick is designed for use as a backup on interior masonry?

    a. Grade NW

    b. Grade AW

    c. Grade MW

    d. Grade SW

22. _____ units are reduced-thickness units, or fire clay, used in veneer applications.

    a. Glazed brick

    b. Thin brick veneer

    c. Green bricks

    d. Paving brick

23. A modular unit is based on a measurement of ___ inch(es).

    a. 1

    b. 3

    c. 4

    d. 5

24. In contrast to a lightweight block, a typical 8" × 16" aggregate and sand block weighs about ___ pounds.

    a. 20

    b. 30

    c. 40

    d. 50

25. Which of the following high-compressive strength mortars is to be used for unreinforced masonry below grade and in contact with the earth?

    a. Type M

    b. Type S

    c. Type N

    d. Type K

Notes

Notes

PART THIRTEEN
Carpentry

Questions are based on general knowledge of carpentry.

The following quiz is based on topics and issues of basic carpentry. As opposed to questions based on code, these address the why and how. Refer to your candidate information bulletin for the approved reference(s) that relate to masonry. The answers to these questions will *not* be answered in the code book.

Questions in this quiz cover the following subject matter:

- Tools
- Site work
- Framing
- Concrete
- Plans

1. The location of a residential home as it should be constructed on the building site is specified on the ____ plan.

   a. floor

   b. plot

   c. plat

   d. framing

2. Which of the following floor framing members supports the heaviest load of attached horizontal members?

   a. Joists

   b. Ledger strips

   c. Lintels

   d. Girders

3. What is the standard height for an interior door for residential homes?

   a. 7 feet

   b. 7 feet 2 inches

   c. 6 feet

   d. 6 feet 8 inches

4. The term used to describe the inside of a door opening during the framing stage and prior to the unit installation is:

   a. Net opening

   b. Door opening

   c. Rough opening

   d. Trimmer dimensions

5. What is used to position a leveling instrument directly over a given point during surveying or foundation construction?

   a. Benchmark

   b. Plumb bob

   c. Laser level

   d. Point indicator

6. Which of the following is a common squaring method used to ensure an angle is 90 degrees?

   a. 2-3-6

   b. 2-4-6

   c. 3-4-5

   d. 6-8-9

7. When joists are parallel to a component, such as in a shower or garden tub, the joists should be:

   a. Doubled

   b. Blocked

   c. Braced

   d. Engineered

8. A common field test to verify a rectangular building is properly square at each corner is to:

   a. Make sure the width and length are the same

   b. Check the corners with a building square

   c. Measure the diagonal corners, and if they are equal, the building is square

   d. Check at least one corner using the 3,4,5 system

Notes

9. What are the three members of a balustrade?

   a. Newels, handrails, and balusters

   b. Treads, risers, and handrails

   c. Stringers, handrails, and newels

   d. Balusters, handrails, and risers

10. What type of saw is used to shape the mitered edges of molding?

   a. Hack

   b. Coping

   c. Jig

   d. Skill

11. What type of saw blade is used for cutting a groove in wood?

   a. Dado

   b. Flat

   c. Cross-cut

   d. Kerf

12. The point at which contractors and surveyors begin an excavation is the _____ .

   a. level

   b. datum

   c. hub

   d. bench mark

13. Batter boards are typically placed _____ .

   a. at the very edge of the foundation lines

   b. 2 feet beyond the edge of the foundation lines

   c. 4 feet beyond the edge of the foundation lines

   d. inside the edge of the foundation lines

14. What size header would be required for a 36" door in a typical wood frame construction application?

    a. 37"

    b. 38½"

    c. 41½"

    d. 43"

15. When pouring a 6" slab, how many square feet will a cubic yard of concrete cover?

    a. 54 square feet

    b. 66 square feet

    c. 76 square feet

    d. 81 square feet

16. To prevent termites from entering a wood sill, use _____ .

    a. silicone

    b. metal flashing

    c. caulking

    d. termite preventor

17. A _____ is used to produce a rounded edge on a concrete slab.

    a. bull float

    b. edger

    c. screed

    d. spade

Notes

Notes

18. What is the recommended maximum height of a stepped footing?

    a. 1 foot

    b. 2 feet

    c. 3 feet

    d. 4 feet

19. Plans are to be scaled using ¼ inch = 1 foot. On the plans, how many inches would represent a 34-foot wall?

    a. 4.3 inches

    b. 8.5 inches

    c. 9.6 inches

    d. 12.6 inches

20. If a sheet of plywood is stamped with 36/16, this indicates that the panel is rated for:

    a. Roof framing spaced 16 inches on center and floor framing spaced 36 inches on center

    b. Roof framing spaced 36 inches on center and floor framing spaced 16 inches on center

    c. Roof framing spaced 36 inches on center and floor framing spaced 36 inches on center

    d. Roof framing spaced 16 inches on center and floor framing spaced 16 inches on center

21. What is constructed for the purpose of diverting water around a chimney?

    a. Cricket

    b. Flashing

    c. Valley strips

    d. Diverter

22. Which of the following members is installed to prevent opposing rafters from spreading apart?

    a. Collar tie

    b. Bracing

    c. Rafter jack

    d. Rafter cripples

23. Each of the following statements is *true* about architectural grade glulams *except*:

    a. It is much stronger than a framing grade glulam

    b. All exposed edges are surfaced

    c. Appearance is important but some voids are allowed

    d. Exposed edges are slightly rounded

24. Which of the following statements about cross bridging is *true*?

    a. It is an effective means of bracing

    b. It requires less material

    c. It is not generally required by codes

    d. All of the above

25. The type of drawing to use when precise information is needed about a complex portion of the building is a _____ .

    a. rendering

    b. schedule

    c. detail

    d. section view

Notes

Notes

PART FOURTEEN

Project Management

Questions are based on business and project management.

The following questions are typical of those found in the project management section of most state contractor licensing exams. While many state requirements include separate exams to cover this content, several states incorporate these questions intermittently with the trade questions. Be careful to read your candidate information bulletin to determine whether your exam covers this subject matter and to determine the approved reference material.

Questions in this quiz cover the following subject matter:

- Accounting
- Scheduling
- Fair Labor Standards Act (FLSA)
- Business organization
- Payroll
- Federal Unemployment Tax (FUTA)

Project Management Practice Exam 1

1. Each of the following is regulated by the FLSA *except*:

   a. Vacation

   b. Time and a half for hours over 40

   c. Minimum wage

   d. Compliance with child labor laws

2. What amount does an employer pay for Social Security and Medicare?

   a. 6.2%

   b. 1.45%

   c. 7.65%

   d. 5.4% on the first $7,000

3. Two brothers are planning to start a plumbing contracting company together. The oldest brother will be the primary financial backer and wants no day-to-day involvement in the operations. What type of business should the brothers form?

   a. General partnership

   b. Proprietorship

   c. Limited partnership

   d. Corporation

4. The best source for projecting future project and company overhead percentage rates is:

   a. Historical data

   b. Bid documents

   c. Unit pricing

   d. Job cost records

5. Which of the following insurance types is most widely used for construction projects?

   a. Building operations

   b. Liability

   c. All-risk builder's-risk

   d. Catastrophe

6. Which of the following would be considered project overhead?

   a. Yellow page advertising

   b. Secretary's salary

   c. Electric bill for the office

   d. Electric bill for the project office

7. Yellow Duck Plumbing Company has 21 employees. According to OSHA, this company must keep a record for _____ years after each recorded accident while on the job site for each of its employees.

   a. 3

   b. 4

   c. 5

   d. 6

8. What entity continues to exist beyond the death of one of the principals?

   a. Corporation

   b. Partnership

   c. Sole proprietorship

   d. Limited partnership

Notes

Notes

9. Employers are not allowed to deduct _____ from employee's wages.

   a. Social Security

   b. Medicare

   c. FUTA

   d. Federal Income Tax

10. Which of the following is a *disadvantage* of setting up a general partnership?

    a. Single taxation

    b. Easy setup

    c. Unlimited personal liability

    d. The inability to pool financial resources

11. Which of the following would *not* constitute a material breach of contract?

    a. Building on the wrong site

    b. A refusal to work

    c. The owner refusing to pay

    d. A delay in the work

12. What form of business organization offers contractors the most protection against personal liability?

    a. Corporation

    b. Partnership

    c. Joint venture

    d. Sole proprietorship

13. *Given:* A carpenter drops a hammer from a roof and injures a curious bystander. Which type of insurance would cover the victim's medical costs?

    a. Builder's risk

    b. Named-peril risk

    c. Liability

    d. Risk management

14. The simplest form of business ownership is _____ .

    a. corporation

    b. partnership

    c. limited partnership

    d. sole proprietorship

15. What form is used to report FUTA?

    a. W-3

    b. 940

    c. W-2

    d. 1099

16. A large plumbing contractor maintains a central office and has crews working at several projects. Which of the following is considered a company overhead cost?

    a. Superintendent salary

    b. Temporary barricades

    c. Site cleanup

    d. Advertising

17. Typically, the only party with whom subcontractors and their employees have a contractual relationship is the _____ .

    a. owner's representative

    b. architect

    c. owner

    d. contractor

18. The purpose of a journal is to:

    a. Post transactions on a weekly basis

    b. Post transactions on a daily basis

    c. Ensure proper job costing

    d. Balance the checkbook

19. What is the latest date an employer can file a FUTA return (Form 940)? *Note*: Assume no time extensions are taken and the balance due is greater than $100.00.

    a. December 1

    b. December 31

    c. January 1

    d. January 31

20. Employee records regarding pay shall be kept for at least _____ year(s).

    a. 1

    b. 2

    c. 3

    d. 4

21. Which of the following is the proper formula for determining assets?

    a. Liabilities + Owner's equity

    b. Liabilities − Owner's equity

    c. Owner's equity − Liabilities

    d. Assets − Liabilities

22. Which of the following components make a change order complete?

    a. The change to be made, the charge, and the effect on time

    b. The owner's/contractor's address and the change order number

    c. The legal description and the contractor's license number

    d. The contractor's license number, the change order number, and the architect's signature

23. The primary purpose of a written contract is to:

    a. Define rights and obligations of each party

    b. Standardize all requirements of work performance

    c. Condense construction requirements

    d. All of the above

24. The term used in scheduling that makes reference to the TIME a project can be delayed without extending the project completion:

    a. Float time

    b. Lag time

    c. Critical task

    d. Noncritical task

Notes

25. Which of the following terms is used to describe withholding a percentage of a progress payment pending completion to a certain point for the purpose of covering defects or faulty installation?

    a. Retainage

    b. Withholding

    c. Holdback

    d. None of the above

Project Management Practice Exam 2

1. A bid bond provides:

   a. Assurance that the bonding company is liable for all bid dis-crepancies

   b. Assurance to the owner or government agency that the con-tractor will honor the bid submitted by entering into a contract in accordance with the terms of the bid

   c. A guarantee to the owner that the project will be completed

   d. A guarantee of both job completion and payment of all labor and materials

2. A truck that is due to be paid off within a year would be consid-ered a(n):

   a. Current liability

   b. Long-term liability

   c. Asset

   d. Other

3. In the event a subcontractor does not carry worker's compensa-tion insurance on his employees, who is responsible for providing the insurance?

   a. Owner

   b. General contractor

   c. State government

   d. No one

4. Which of the following financial statements indicates the financial condition of a company at a given point in time?

   a. Income statement

   b. Cash flow statement

   c. Balance sheet

   d. Funds flow statement

5. A first aid kit is checked before it is sent to the job and is then checked:

   a. Every week

   b. Biweekly

   c. Monthly

   d. Quarterly

6. Which of the following would be classified as a long-term debt?

   a. Current annuity of long-term debt

   b. Mortgage on a company office building

   c. Accounts payable

   d. Accumulated depreciation

7. What is another name for owner's equity?

   a. Operating capital

   b. Liquid assets

   c. Net worth

   d. Liability

8. Which of the following statements defines *net working capital*?

   a. The amount that current assets exceed the current liabilities

   b. The amount listed as an item overhead expense

   c. The difference between direct cost and the cost of sales

   d. Current liabilities less overhead expense

9. When two companies enter into a project as a joint venture, what must occur if one company does not have a contractor's license?

   a. The company with the license will manage

   b. A license in the name of the joint venture shall be obtained

   c. This is permitted as long as at least one firm has a license

   d. Other

10. Based on an accrual basis of accounting, earnings that have not been received will be displayed as:

    a. Current assets

    b. Fixed assets

    c. Current, liabilities

    d. Owner's equities

11. The required quality of a project is described in the _____ .

    a. budget

    b. specifications

    c. plans

    d. general conditions

12. Current assets are assets that can be converted into cash within:

    a. 1 year

    b. 6 months

    c. 1 month

    d. 2 years

13. A form of business organization that has a tax structure similar to a partnership but is organized as a corporation is a _____ .

    a. limited partnership

    b. subchapter s corporation

    c. general partnership

    d. nonprofit organization

Notes

14. Regarding contracts, which of the following statements is *not true*?

    a. Words prevail over numbers

    b. General clauses prevail over special clauses

    c. When properly documented, handwritten agreements prevail over typed provisions

    d. A contract deemed ambiguous is interpreted against the party that provided it

15. Which of the following work-related injuries does *not* require the employer to report to OSHA?

    a. A broken wrist

    b. An x-ray

    c. Cancer resulting from long-term exposure to work-related chemicals

    d. A sprain that removed an employee from work for 2 days

16. What is a benefit of good inventory control?

    a. Better for safety

    b. Better preparation for increase in business

    c. Reduced time and expense in handling materials

    d. Ability to ship faster to off-site storage

17. Periodic payments made by the owner to the contractor over a long project are called:

    a. Progress payments

    b. Acceleration payments

    c. Performance payments

    d. Provisional payments

18. What accounting method records income when cash is received?

    a. Accrual basis

    b. Cash basis

    c. Single entry

    d. Double entry

19. What portion of the worker's compensation insurance premium can an employer charge to an employee?

    a. A maximum of 2% of the employee's weekly gross earnings

    b. No deduction is allowed

    c. Up to 50% of the premium

    d. A flat fee of $25 per week

20. If an employee working for a corporation and the licensing examination for the corporation, who owns the license when it is issued?

    a. The employee

    b. The stockholders

    c. The president of the corporation

    d. The corporation itself

21. An employer must maintain records about work-related injuries for _____ year(s).

    a. 3

    b. 5

    c. 1

    d. 7

Notes

22. Which of the following statements is *not true* regarding a bar chart method of scheduling?

    a. It shows the start and finish times for each task

    b. It does not show how tasks relate to each other

    c. It is most common scheduling system found on the job site

    d. All are true

23. When a clause in a contract calls for liquidation damages, it is referring to:

    a. A payment per day for finishing after the contract deadline

    b. A payment due to flooding

    c. An expedient payment for improper work

    d. None of the above

24. Which of the following statements is *true* about partnerships?

    a. Each partner may cause the other to be jointly liable for any action incurred during normal business

    b. Each partner is personally liable for his actions only

    c. Each partner may sell his share of the partnership without the consent of the other partners

    d. A general partner may discharge himself from all liability by declaring himself to be a limited partner

25. Which of the following acts requires payment of prevailing wage rates and fringe benefits on federally assisted construction?

    a. Davis-Bacon Act

    b. Walsh-Healey Public Contracts Act

    c. Wage Garnishment Law

    d. Fair Labor Standards Act

PART FIFTEEN
OSHA

Questions are based on safety issues as they relate to OSHA.

The following section is based on OSHA 1926, Construction Industry Regulations. If the OSHA book is an approved reference material listed for your exam, these questions represent the types of questions that will be covered.

Questions in this quiz cover the following subject matter:

- Scaffolding
- Ladders
- Fall protection
- Excavation
- Storage
- Tools

OSHA Practice Exam 1

1. The maximum slope for an 8 feet excavation in Type B soil is:

    a. ¾:1

    b. ½:1

    c. 1:1

    d. 2:1

2. The minimum perpendicular clearance between fixed ladder rungs, cleats, and steps and any obstruction behind the ladder is _____ inches.

    a. 6

    b. 7

    c. 8

    d. 9

3. The use of electrical cords for hoisting or lowering shall:

    a. Be allowed

    b. Not be allowed

    c. Only be permitted during daylight hours

    d. Not be permitted during dusk or evening hours

4. When necessary to require employees to work in a trench, the excavated materials shall be placed a minimum of ____ away from the edge of the trench.

    a. 0'

    b. 1'

    c. 2'

    d. 3'

5. Eye protection near dangerous working conditions is:

   a. Required at the employee's cost.

   b. Required at the employer's cost.

   c. Not required, except by union regulations

   d. Not required

6. Who is responsible for the fire safety and prevention on a job site through completion of a project?

   a. OSHA

   b. Employer

   c. Fire marshall

   d. Building inspector

7. When employees are required to be in trenches of 4 feet deep or more, an adequate means of exit such as a ladder or steps shall be provided and located so as to require no more than ___ feet of lateral travel.

   a. 10

   b. 15

   c. 25

   d. 30

8. How much water is allowed in an excavation before work must cease?

   a. 3" or less at the edge

   b. 2" or less at the edge

   c. 1" or less at the edge

   d. None

Notes

9. The minimum illumination in active general construction storage areas is:

    a. 3 footcandles

    b. 5 footcandles

    c. 10 footcandles

    d. 30 footcandles

10. After being sent to the job site, a first aid kit is checked _____ .

    a. weekly

    b. biweekly

    c. monthly

    d. quarterly

11. OSHA requires a ladder in every excavation exceeding:

    a. 5 feet

    b. 3 feet

    c. 4 feet

    d. 6 feet

12. Compressed air shall not be used for cleaning purposes, except where reduced to less than _____ psi.

    a. 10

    b. 20

    c. 30

    d. 70

13. Sand is considered what soil type?

    a. Type A

    b. Type B

    c. Type C

    d. Type D

14. According to OSHA, when employees are exposed to inhalation, ingestion, skin absorption, or contact with any substance at concentrations above the _____ , they shall be evacuated unless specific forms of personnel protective equipment or devices are provided.

    a. threshold limit value

    b. PPM mandate

    c. PEL levels

    d. oxygen deprivation level

15. According to OSHA, the height at which fall protection is required is ____ feet.

    a. 6

    b. 8

    c. 10

    d. 12

16. Circular saws shall be equipped with a _____ switch.

    a. position on-off

    b. momentary on-off

    c. lock-on control

    d. constant pressure

17. Combustible debris should be removed from a construction job site _____ .

    a. when the job is complete

    b. only in the event of hazardous fire conditions

    c. periodically by any means

    d. at regular intervals

Notes

18. When using a stepladder:

    a. The top step can be used

    b. The cross bracing on the back can be used as a step

    c. Both hands must be used on the sides

    d. It should be inspected by a competent person for visible defects on a periodic basis

19. When employees are required to be in trenches of 4 feet deep or more, an adequate means of egress shall be provided and located so as to require not more lateral travel than _____ feet.

    a. 10

    b. 12

    c. 15

    d. 25

20. The maximum allowable slope for excavations less than 20 feet in depth for Type C soil is:

    a. ¾ horizontal to 1 vertical

    b. 1 horizontal to 1 vertical

    c. 1½ horizontal to 1 vertical

    d. 2 horizontal to 1 vertical

21. A 10-foot-deep trench is to be dug in granular sand. What is the maximum allowable bank slope?

    a. 1:1

    b. 1½:1

    c. ½:1

    d. ¼:1

22. When using a portable ladder for reaching an elevated surface or for ingress or egress of an excavation, the ladder must extend to what minimum height above the surface?

    a. 2.5 feet

    b. 3 feet

    c. 3.5 feet

    d. 4 feet

23. _____ soil has a maximum allowable slope of 1:1.

    a. Type A

    b. Type B

    c. Type C

    d. Type D

24. Each of the following is a soil type *except*:

    a. Gravel

    b. Sediment

    c. Sand

    d. Clay

25. OSHA requires the employer to provide _____ for all employees/workers in all operations where there is an exposure to hazardous conditions.

    a. no-smoking areas

    b. an opportunity to change positions

    c. new tools

    d. hardhats

OSHA Practice Exam 2

1. What is the *maximum* amount of hours an employee can work if exposed to a D.B.A level of 102?

   a. 1 hour

   b. 1.5 hours

   c. 2 hours

   d. 3 hours

2. When intermediate vertical members such as balusters used between posts are positioned, the space between them shall not exceed ____ inches.

   a. 10

   b. 15

   c. 19

   d. 20

3. Combustible material stored outdoors shall be at least how many feet away from the structure?

   a. 20 feet

   b. 50 feet

   c. 10 feet

   d. 75 feet

4. How often should powder-actuated tools be tested?

   a. Once a week

   b. Per manufacturer's requirements

   c. Each day after use

   d. Each day before use

5. Safety belt lanyard hardware shall be able to withstand a tensile loading strength of up to _____ pounds before cracking, breaking, or deforming.

    a. 1,000

    b. 3,000

    c. 4,000

    d. 5,000

6. Hooks on an attachable ladder shall be positioned so that the bottom rung is not more than ____ inches above the scaffold supporting level.

    a. 10

    b. 18

    c. 20

    d. 24

7. Stacks of lumber that are to be handled manually shall not be stacked more than ____ feet.

    a. 10

    b. 15

    c. 16

    d. 20

8. When debris is dropped through holes in the floor without the use of chutes, the area onto which the material dropped shall be completely enclosed with barricades and located not less than _____ feet back from the projected edge of the opening above.

    a. 2

    b. 4

    c. 6

    d. 8

Notes

9. All equipment used in site clearing shall be equipped with a roll-over canopy structure of not less than ___ steel plate or ____ wire mesh with openings of no more than 1 inch.

   a. ¼"

   b. ⅛"–¼"

   c. ¼"–⅛"

   d. ½"–¾"

10. Integral prefabricated scaffold access frames shall be provided with rest platforms at ____ foot intervals.

    a. 15

    b. 25

    c. 35

    d. 45

11. What is the required rating for fire extinguishers placed on job sites where more than 5 gallons of flammable liquids are used?

    a. 5A

    b. 5B

    c. 10A

    d. 10B

12. A safety belt lanyard shall have a length allowing a maximum fall of ___ feet.

    a. 2

    b. 4

    c. 6

    d. 8

13. When bricks are stored on a job site, the maximum height they can be stacked without additional requirements is ____ feet.

    a. 5

    b. 6

    c. 7

    d. 8

14. How often should first aid kits be checked?

    a. Monthly

    b. Weekly

    c. Biweekly

    d. Daily

15. What is the minimum diameter of a safety lanyard?

    a. ¼ inch

    b. ⅜ inch

    c. ½ inch

    d. ¾ inch

16. Fiber rope slings shall have a minimum clear length of rope between eye splices equal to ____ times the rope diameter.

    a. 5

    b. 10

    c. 15

    d. 20

Notes

Notes

17. All handheld circular saws shall be equipped with a:

    a. Positive on-off switch

    b. Momentary contact switch

    c. Constant pressure switch

    d. None of the above

18. When dropping waste more than ____ feet, an enclosed chute of wood or an equivalent shall be used.

    a. 10

    b. 15

    c. 20

    d. 25

19. Safety nets shall be provided when workplaces are more than ____ feet above the ground and no other protection is provided.

    a. 10

    b. 15

    c. 20

    d. 25

20. A ladder jack scaffold platform shall not exceed ____ feet in height.

    a. 10

    b. 15

    c. 20

    d. 25

21. A fixed guardrail system, used to protect an opening must be capable of sustaining the pressure exerted by a ____ person if there is a chance of someone falling through it.

    a. 150 lb.

    b. 175 lb.

    c. 200 lb.

    d. 225 lb.

22. Safety nets must extend beyond the work surface by at least ____ feet.

    a. 4

    b. 6

    c. 8

    d. 10

23. The minimum distance between a scaffold and a 300-volt power insulated line is ___ feet.

    a. 3

    b. 6

    c. 9

    d. 10

24. When using a ladder, it must extend at least ___ feet above the surface to which it is providing access.

    a. 1

    b. 2

    c. 3

    d. 4

Notes

Notes

25. Stairway-type ladders shall be provided with a rest platform at ____ foot *maximum* vertical intervals.

    a. 8

    b. 10

    c. 12

    d. 14

PART SIXTEEN

Comprehensive Final

Questions are based on general knowledge of concrete.

The following comprehensive final is based on questions from the preceding chapters. This exercise will prove most beneficial to exam candidates that are preparing to take an exam with multiple approved reference books. By taking questions from a wide variety of subjects, the importance of reading the question, deciding what chapter of the code to locate, or whether or not the question can be answered in the code book becomes apparent.

Certain words will hint that the question can be answered from the code book. These words include:

- Minimum
- Maximum
- Must be
- Shall be
- Required

Questions that are based on theory and general knowledge can be identified by words such as:

- Recommended
- Suggested
- What is used
- Typically

Notes

1. Utility-grade studs cannot be spaced more than _____ inches on center.

   a. 12

   b. 16

   c. 20

   d. 24

2. Foundation plates or sills should be attached to concrete foundation with bolts:

   a. ½ inches in diameter with a penetration of at least 7 inches

   b. ¾ inches in diameter with a penetration of at least 7 inches

   c. ½ inches in diameter with a penetration of at least 6 inches

   d. ¾ inches in diameter with a penetration of at least 6 inches

3. The type of mortar best suited for use below grade is:

   a. Type K

   b. Type M

   c. Type N

   d. Type S

4. The building permit must be:

   a. Displayed in the contractor's office

   b. Posted on the job site

   c. Provided to the inspector when requested

   d. Posted in the building official's office

5. A brick cut lengthwise across the end is called a _____ closure.

  a. queen

  b. king

  c. bat or half

  d. split

6. The best source for projecting future project and company overhead percentage rates is _____ .

  a. historical data

  b. bid documents

  c. unit pricing

  d. job cost records

7. Exterior air intake shall not be located in any of the following areas *except*:

  a. A garage

  b. Outside a dwelling

  c. A basement

  d. An attic

8. Which of the following is a consideration when applying tile?

  a. Climate conditions

  b. Roof slope

  c. Type of tile

  d. All of the above

Notes

Notes

9. The use of electrical cords for hoisting or lowering shall:

   a. Be allowed

   b. Not be allowed

   c. Only be permitted during daylight hours

   d. Not be permitted during dusk or evening hours

10. Normal concrete is designed to reach its intended compressive strength within _____ days.

    a. 7

    b. 14

    c. 21

    d. 28

11. For ceilings not attached to rafters with a roof live load of 20 psf and a dead load of 10 psf, the maximum span allowed for a 2 × 8 rafter to be spaced 16 inches on center is _____. *Note:* Material is SYP #2.

    a. 18 feet 6 inches

    b. 19 feet 5 inches

    c. 14 feet 4 inches

    d. 19 feet 3 inches

12. What is the minimum depth of a masonry chimney firebox?

    a. 20 inches

    b. 24 inches

    c. 36 inches

    d. 301

13. The minimum uniformly distributed live load for a room with an intended use of sleeping is designated as _____ psf.

    a. 10

    b. 20

    c. 30

    d. 50

14. For the purpose of the code, a vehicle covering is considered a carport if it is open on:

    a. At least one side other than the entry

    b. All four sides

    c. At least two sides

    d. At least one side and not adjacent to the structure

15. For each inch of reduced slump in concrete, the pounds per square inch is reduced by:

    a. 100 psi

    b. 125 psi

    c. 150 psi

    d. 200 psi

16. What is another name for owner's equity?

    a. Operating capital

    b. Liquid assets

    c. Net worth

    d. Liability

Notes

17. The net free ventilation area for attics must be at least _____ of the area. *Note:* No exceptions.

    a. $\frac{1}{150}$

    b. $\frac{1}{200}$

    c. $\frac{1}{250}$

    d. $\frac{1}{300}$

18. Wood shakes shall be laid with a minimum side lap not less than _____ inches.

    a. $\frac{3}{4}$

    b. 1

    c. $1\frac{1}{4}$

    d. $1\frac{1}{2}$

19. Calculate the common number of rafters for a building that is 24 feet long with a 1 foot overhang. The slope factor for the building is 1.118.

    a. 14

    b. 15

    c. 16

    d. 17

20. Sand is considered to be what soil type?

    a. Type A

    b. Type B

    c. Type C

    d. Type D

21. A plain masonry hollow wall that is 7 feet tall with 6 feet of unbalanced backfill should be at least _____ inches thick. The soil type is GW.

    a. 6

    b. 8

    c. 10

    d. 12

22. Chimney terminations shall not be less than _____ feet above the highest point where the chimney passes through the roof.

    a. 2

    b. 3

    c. 6

    d. 10

23. A permit is deemed void if work does not begin within _____ days of issuance.

    a. 30

    b. 60

    c. 90

    d. 180

24. The minimum diameter head permitted to be on a fastener for asphalt shingles is _____ inches?

    a. 0.105

    b. 0.375

    c. 0.50

    d. 0.75

25. A smoke-developed index of no more than _____ should be maintained for all wall and ceiling finishes.

    a. 200

    b. 250

    c. 375

    d. 450

26. Beams and girders must bear at least _____ inch(es) on a masonry column.

    a. 1

    b. 1½

    c. 1¾

    d. 3

27. Vapor retarder is not required between the base course/subgrade and a concrete floor slab in which of the following circumstances?

    a. Typical driveway and walkways

    b. Unheated accessory structures

    c. For installations approved by the building official based on local site conditions

    d. All of the above

28. Purlins used to support roof loads must be at least:

    a. 2 × 6

    b. One size larger than the rafter it is bracing

    c. No less than one size smaller than the rafter it is bracing

    d. The size of the rafter it is supporting

29. *Given:* A carpenter drops a hammer from a roof and injures a curious bystander. Which type of insurance would cover the victim's medical costs?

    a. Builder's risk

    b. Named-peril risk

    c. Liability

    d. Risk management

30. How many studs will be needed if a wall is 44 feet long and the spacing is 16 inches on center?

    a. 33

    b. 34

    c. 35

    d. 36

31. What is the nominal dimension of a standard concrete masonry unit (Block)?

    a. $7\frac{5}{8}" \times 7\frac{5}{8}" \times 16"$

    b. $7\frac{5}{8}" \times 8" \times 15\frac{5}{8}"$

    c. $8" \times 7\frac{5}{8}" \times 15\frac{5}{8}"$

    d. $8" \times 8" \times 16"$

32. SPF #2 is to be installed 16 inches on center in a residential living area with a dead load of 20 psf. The length of the room is 14 feet 6 inches. What size floor joist is required?

    a. 2×6

    b. 2×8

    c. 2×10

    d. 2×12

Notes

33. What is the maximum stud spacing if particleboard wall sheathing is to be nailed to the stud walls? *Note:* M-2 Exterior glue grade, ½ inch thickness.

    a. 12 inches on center

    b. 16 inches on center

    c. 24 inches on center

    d. 30 inches on center

34. Particleboard (M-1/M-S grade) wall sheathing nailed to 16 inches on center studs must be at least _____ inch(es) in thickness.

    a. ½

    b. ⅜

    c. ⅝

    d. 1

35. The most common welded-wire fabric used for residential applications has wires spaced _____ inches apart in each direction.

    a. 2

    b. 4

    c. 6

    d. 8

36. Masonry chimneys shall be lined appropriately for the type of appliance connected. Residential-type appliances having a clay flue lining that must comply with the requirements of:

    a. ASTM C 27

    b. ASTM C 199

    c. ASTM C 315

    d. ASTM 1261

37. What is the maximum notch depth allowed for a 2 × 12 nominal floor joist? *Note:* The joist spans 14 feet 6 inches and the notch needs to be made 5 feet from one end.

    a. 1⅞"

    b. 1¾"

    c. 2"

    d. None of the above

38. When single top plates are permitted, the plate must be adequately tied at joints, corners, and intersecting walls by a galvanized steel plate that is nailed to each wall or segment by _____ nails on each side.

    a. Eight 16d

    b. Four 16d

    c. Six 8d

    d. Four 8d

39. A 2 × 6 SYP #2 material is to be used for the ceiling joists with a live load of 10 psf. The joists, spaced 24 inches on center, will be installed to support an attic that will not be used for storage. What is the maximum span?

    a. 11 feet 0 inches

    b. 15 feet 6 inches

    c. 15 feet 11 inches

    d. 17 feet 8 inches

40. Coal-tar saturated organic felt used for built-up roof cover must conform to what standard?

    a. ASTM D 227

    b. ASTM D 450

    c. ASTM D 43

    d. ASTM D 4022

Notes

41. *Given:* The floor of a combustion chamber to the top of the flue is 18 feet. The fireplace opening is 2,200 square inches. What is the minimum cross-sectional area required for a retangular chimney flue?

    a. 125 square inches

    b. 140 square inches

    c. 168 square inches

    d. 214 square inches

42. Hydrated lime that has been formed into putty before packaging is called _____ .

    a. mixed

    b. prehydrated

    c. slaked

    d. preserved

43. _____ soil has a maximum allowable slope of 1½:1.

    a. Type A

    b. Type B

    c. Type C

    d. Type D

44. For an emergency escape opening in a habitable space, the minimum net clear opening width is specified as _____ inches.

    a. 20

    b. 22

    c. 24

    d. 28

45. Handrail height for stairs, when required, should be a minimum _____ inches and a maximum _____ inches, measured vertically from the sloped plane adjoining the tread nosing.

    a. 38/34 (respectively)

    b. 34/36 (respectively)

    c. 34/38 (respectively)

    d. 36/38 (respectively)

46. Gravel or crushed stone drains installed for the purpose of foundation drainage shall extend at least _____ beyond the outside edge of the footing.

    a. 0.5 feet

    b. 1 foot

    c. 1.5 feet

    d. 2 feet

47. A one-story building with a width of 28 feet requires _____ girder(s). *Note:* The span is 10 feet 2 inches.

    a. Four 2 × 12

    b. Four 2 × 10

    c. Four 2 × 8

    d. Three 2 × 10

48. Fasteners for attaching structural sheathing (shear panels) to steel wall framing must be installed with a minimum edge distance of _____ inch.

    a. ¼

    b. ½

    c. ⅜

    d. 1

Notes

49. What species of lumber must be used if 2×6 nominal ceiling joists are to span 14 feet 8 inches and be spaced 16 inches on center? The live load is 20 psf, and the application is uninhabitable attics with limited storage.

    a. Douglas fir-larch

    b. Hem-fir

    c. Southern pine

    d. Spruce-pine-fir

50. All of the following fasteners are used for asphalt shingles *except:*

    a. Galvanized steel

    b. Copper

    c. Aluminum

    d. Nickel

51. Based on a brick factor of 6.5 per square foot, how many bricks will be needed for a wall that is 35 feet long and 12 feet tall?

    a. 2,730

    b. 2,600

    c. 2,450

    d. 2,340

52. Footings for masonry chimneys in areas not subjected to freezing shall be at least _____ inches below finished grade.

    a. 6

    b. 12

    c. 18

    d. 24

53. Chimneys located inside a building shall have minimum airspace combustibles of _____ inches.

    a. 2

    b. 4

    c. 6

    d. 12

54. The space between the end of the rebar and the form is referred to as _____ .

    a. cover

    b. bond

    c. space

    d. void

55. _____ units are reduced-thickness units, or fire clay, used in veneer applications.

    a. Glazed brick

    b. Thin brick veneer

    c. Green bricks

    d. Paving bricks

56. ABC Contractors Company has 21 employees. According to OSHA, this contractor must keep a record for _____ years after each recorded accident while on the job site for each one of its employees.

    a. 3

    b. 4

    c. 5

    d. 6

Notes

57. Combustible debris should be removed from a construction job site _____ .

    a. when the job is complete

    b. only in the event of hazardous fire conditions

    c. periodically and by any means

    d. at regular intervals

58. Untreated wood joists should not be installed closer than _____ inches.

    a. 12

    b. 14

    c. 16

    d. 18

59. Where masonry veneer is used, concrete foundation walls are required to extend _____ inches above the finished grade adjacent to the foundation walls at all points.

    a. 2

    b. 4

    c. 6

    d. 8

60. Sheathing with a span rating of 24/0 to be used for a subfloor can be installed with a maximum span of _____ inches.

    a. 0

    b. 16

    c. 24

    d. 32

61. Buildings must be provided with exterior and interior braced wall lines. The spacing of the braced wall lines shall not exceed _____ feet in both the longitudinal and transverse directions in each story. *Note:* No exceptions apply.

    a. 10

    b. 25

    c. 35

    d. 50

62. The type of nails that can be substituted for 16d nails when fastening rafter connections is:

    a. 10d

    b. 12d

    c. 40d box

    d. No substitutions provided by code

63. On roofs sloped 4:12 or greater, underlayment shall be at least one layer applied shingle fashion, starting from the eaves and lapped _____ inch(es).

    a. 1

    b. 2

    c. 4

    d. 5

64. What is the absolute minimum thickness of a heart extension?

    a. ⅜"

    b. 2"

    c. 4"

    d. 8"

65. With regard to masonry units, a *nominal* dimension refers to:

    a. Actual dimension

    b. Standard brick sizes

    c. Nominal brick plus the mortar joint

    d. Actual brick size plus the mortar joint

66. A 10 feet × 10 feet habitable room should have an aggregate glazed area of _____ square feet to provide natural lighting.

    a. 5

    b. 6

    c. 7

    d. 8

67. Pier and curtain wall foundations are permitted to support light frame construction of not more than two stories. All of the following requirements must be met by code *except:*

    a. The load-bearing wall shall be placed on continuous concrete footings that are placed integrally with the exterior wall

    b. The thickness of the load-bearing wall shall not be less than 3⅜" nominal

    c. The maximum height of a 4" load-bearing masonry foundation wall supporting wood framed walls shall be more than 4' in height

    d. The unbalanced fill for 4" foundation walls shall not exceed 24" for solid masonry.

68. A hold to be bored into a floor joist for the purpose of running a plumbing pipe or electrical conduit:

    a. Must not exceed one third of the depth of the joist

    b. Must not exceed one sixth of the depth of the joist

    c. Cannot be bored in the middle third of the joist

    d. Is not allowed unless approved by a qualified engineer

69. Interior partitions (nonbearing) may be constructed with at least a
    _____ top plate.

    a. double top plate

    b. single top plate

    c. double top plate, if wall height exceeds 10 feet

    d. None of the above

70. What is the maximum span of a 2 × 6 SPF #2 ceiling joist to be
    spaced 12 inches on center? The attic will not be utilized in any
    way. Dead load is 5 psf, and the live load is 10 psf.

    a. 18 feet 8 inches

    b. 14 feet 9 inches

    c. 16 feet 3 inches

    d. 13 feet 9 inches

71. Mineral-surfaced roll roofing shall not be applied on roof slopes
    with less than _____ slope.

    a. 2%

    b. 8%

    c. 12%

    d. 33%

72. The upper edge of a chimney cleanout shall be located at least
    _____ inches below the lowest chimney inlet opening.

    a. 2

    b. 4

    c. 6

    d. 8

73. Concrete is made up of how many parts?

    a. 1

    b. 2

    c. 3

    d. 4

74. How many 8 inches × 16 inches CMUs will be needed for a wall that is 45 feet long and 12 feet tall. The total square feet of openings is 200.

    a. 340

    b. 383

    c. 540

    d. 608

75. A _____ masonry wall is constructed with two wythes that can react independently of one another and are separated by a continuous airspace of at least 2 inch (typical).

    a. veneer

    b. hollow

    c. cavity

    d. composite

76. Which of the following statements is *true* regarding site plans submitted for demolition projects?

    a. Site plans are not required for structures to be removed

    b. Site plans must show the construction to be demolished

    c. Site plans must show the size and location of existing construction that will remain

    d. Both b and c

77. A guardrail system is required for protection when a porch or balcony is located more than _____ inches above the grade or floor below.

    a. 24

    b. 30

    c. 36

    d. 40

78. What size and type of gypsum board is required to separate a garage from a habitable room located above it?

    a. ½ inch Type X

    b. ⅝ inch Type X

    c. ¾ inch Type X

    d. 3/16 inch Type X

79. All permanent supports are not required to be protected from frost. Which of the following is required to have frost protection?

    a. A freestanding building of more than 400 square feet

    b. A freestanding building of 400 square feet or less with an eave 10 feet in height.

    c. A freestanding building 400 square feet or less with an eave 8 feet in height

    d. A freestanding building of 300 square feet with an eave 8 feet in height

80. In residential structures, when there is usable space above and below the concealed space of a floor/ceiling assembly, draftstops must be installed so that the area of the concealed space does not exceed _____ square feet.

    a. 500

    b. 800

    c. 1,000

    d. 1,200

Notes

81. When joists exceed a nominal 2 inches × 12 inches (6:1 depth-to-thickness ratio), bridging must be installed for each _____ feet of span.

    a. 4

    b. 8

    c. 12

    d. 16

82. What is the maximum spacing of a 2 × 6 for a bearing wall to support a roof and ceiling only?

    a. 16 inches on center

    b. 20 inches on center

    c. 24 inches on center

    d. 28 inches on center

83. A notch must be placed in a ceiling joist. The nominal size of the joist is 2 × 10 and spans 12 feet 8 inches. What is the maximum notch that can be placed on the top, 2 feet from the exterior wall?

    a. 1.67"

    b. 1.54"

    c. 1.33"

    d. 0

84. Clay tile shall be installed on a roof with a minimum slope of:

    a. 2½:12

    b. 3:12

    c. 4:12

    d. 5:12

85. When employees are required to be in trenches of 4 feet deep or more, an adequate means of exit such as a ladder or steps shall be provided and located so as to require no more than _____ feet of lateral travel.

   a. 10

   b. 15

   c. 25

   d. 30

86. Footings for masonry fireplaces shall extend at least _____ inches beyond the face of the fireplace.

   a. 4

   b. 6

   c. 10

   d. 12

87. The most common type of reinforcement used for the installation of footings is _____ .

   a. deformed steel

   b. smooth steel

   c. rigid steel

   d. #3 smooth steel

88. Which of the following types of brick bond patterns would be considered the weakest?

   a. Flemish

   b. Running

   c. Stack

   d. English

Notes

Notes

89. For ramps intended for use other than specifically for means of egress, the maximum allowable slope is _____ .

    a. 8%

    b. 10%

    c. 10.5%

    d. 12.5%

90. Calculate the board feet of lumber for 12 studs (2 × 4) that are 9 feet in length.

    a. 72 board feet

    b. 78 board feet

    c. 80 board feet

    d. 97 board feet

91. All rescue openings are required to have at least a net clear opening of _____ square feet unless situated on a grade level.

    a. 5.7

    b. 5.0

    c. 4.7

    d. None of the above

92. Joist framing from opposite sides of a beam or girder shall be lapped at least _____ inches.

    a. 2

    b. 3

    c. 4

    d. 6

93. A hole may not be bored closer than _____ inch to the edge of a stud located in an exterior load-bearing framed wall.

    a. ¼

    b. ½

    c. ¾

    d. ⅝

94. A hole must be bored in a ceiling joist and needs to be placed near the edge. What is the closest the hole can be placed to the edge?

    a. 1"

    b. 2"

    c. 2½"

    d. 3"

95. Current assets are assets that can be converted into cash within:

    a. 1 year

    b. 6 months

    c. 1 month

    d. 2 years

96. Which of the following types of wood shakes is *not* approved for grade No. 1?

    a. 18" shakes of naturally durable wood with 7.5" exposure

    b. 24" taper-sawn shakes of naturally durable wood with 7.5" exposure

    c. 24" taper-sawn shakes of naturally durable wood with 10" exposure

    d. 18" preservative-treated taper shakes of Southern yellow pine with 7.5" exposure

97. A chimney shall not be corbelled more than _____ of the chimney's wall thickness from a wall.

    a. 0.25

    b. 0.375

    c. 0.5

    d. 0.75

98. Typical rebar comes in _____ foot lengths that can be cut or bent on the job site according to needs specified.

    a. 10

    b. 15

    c. 20

    d. 25

99. What grade of concrete brick is required when the structure will be exposed to severe frost?

    a. Grade N

    b. Grade S

    c. Grade M

    d. Grade I

100. A sprayed polyurethane foam roof requires a protective coating to be applied no later than _____ after foam application.

    a. 2 hours

    b. 12 hours

    c. 72 hours

    d. can be applied immediately

PART SEVENTEEN
Answer Keys

Part 1: Math Concepts Answer Key

Converting Inches to Feet Answers/Solutions:

1. (c) $14 \div 12 = 1.16666666$. Round by identifying the rounding digit as the second number. The digit to the right of the rounding digit is greater than 5; therefore, the rounding digit is increased by 1, and the remaining digits are dropped.

2. (b) $8 \div 12 = 0.6666666$. Round by identifying the rounding digit as the second number. The digit to the right of the rounding digit is greater than 5; therefore, the rounding digit is increased by 1, and the remaining digits are dropped.

3. (a) $3 \div 12 = 0.25$. Round by identifying the rounding digit as the second number. There are no numbers to the right of the rounding digit, so you are finished.

Applying the Concepts of Rounding and Converting Inches to Feet Answers/Solutions:

1. (b) $8 \div 12 = 0.66666$. Round by identifying the rounding digit as the second number. The digit to the right of the rounding digit is greater than 5; therefore, the rounding digit is increased by 1, and the remaining digits are dropped.

2. (b) $14 \div 12 = 1.16666666$. Round by identifying the rounding digit as the second number. The digit to the right of the rounding digit is greater than 5; therefore, the rounding digit is increased by 1, and the remaining digits are dropped. The total of 14" is derived from adding the space to support a 4" brick, an 8" block, and the 2" of air separating the two units.

3. (c) $21 \div 12 = 1.75$. Round by identifying the rounding digit as the second number. There are no numbers to the right of the rounding digit, so you are finished. The 21" were derived from the information provided. The dimension (to the right) from the grade to the bottom of the slab is denoted as 24". The slab thickness is 5". Add the two together for a total of 29". Next, subtract the 8" brick ledge. The remainder is 21".

4. (b) $29 \div 12 = 2.416666$.

5. (d) $14.67 - 1.17 - 1.17 = 12.33$. Begin by converting inches to feet. 14'8" is converted to 14.67 after dividing the 8 inches by 12 to equal 0.67 (don't forget to round to the nearest hundreth). Next, determine the thickness of the wall by adding the brick, block and airspace. This totals 14 inches. Now convert the 14 inches to feet by diving by 12. 14 divided by 12 = 1.1666666. Rounded to the

nearest hundredth, the thickness of the wall is 1.17. Now, subtract the thicknessof the wall from the overall dimension of the building. $14.67 - 1.17 - 1.17 = 12.33$

6. (b) $14.67 - 0.5 - 0.5 = 13.67$. Begin by converting inches to feet. Subtract the brick and airspace on each side from the overall dimension of the wall.

7. (c) $23 - 1.17 - 1.17 = 20.66$. Subtract the thickness of each wall from the length of the building.

Using Area (Square Feet) to Estimate Sheathing, Drywall, Brick and Block Answers/Solutions:

1. (c) Two walls measure 24' × 8' for a total of 384 square feet. Two walls measure 30' × 8' for a total of 480 square feet. The ceiling measures 24' × 30' for a total of 720 square feet. Add each of these together for a total of 1,584 square feet. The total square feet divided by 32 (which is the square feet of a 4' × 8' piece of drywall) equals 49.5. Round this to 50.

2. (a) 2,875 square feet divided by the area of the material to be installed of 32 square feet equals 89.84. Round to 90 (closest answer and the quantity to order).

3. (b) Half-inch gypsum is specified for walls only. Two walls measure 12' × 8' for a total of 192 square feet. Two walls measure 14' × 8' for a total of 224 square feet. The two totals added together equal 416 square feet. 416 square feet divided by 28 square feet (size of the material) equals 8.67. Round to 9.

4. (b) Two walls measure 20' × 18' for a total of 720 square feet. Two walls measure 32' × 18' for a total of 1,152 square feet. Add each of these together for a total of 2,106 square feet. The total square feet multiplied by 1.125 equals 2,106.

5. (c) Determine the brick multiplier by calculating $7 \times 2.75 = 19.25$ and dividing this into 144 square inches. The multiplier is 7.4805. The square feet of the area is $10 \times 20 = 200 + 10\%$ (waste) for a total of 220 square feet: $220 \times 7.4805 = 1,645.71$, which rounds to 1,646.

6. (a) $32 \times 8 = 256$ square feet of wall area. Subtract the 28 square feet of openings = 228 square feet. Determine the block by multiplying by $1.125 = 256.5$, which rounds to 257. This is the only choice with 257 block; therefore, calculating brick is not necessary. To determine the quantity of brick, begin with determining the multiplier: $2.5 \times 8 = 20$ inches. Divide 144 (square foot) by 20 inches, which equals 7.2. Area of $228 \times 7.2 = 1641.6$ brick (rounds to 1,642).

Notes

Calculate Square Yards Answers/Solutions

1. (a) $20 \times 45 = 900$ square feet. Divide the area (square feet) of 900 by 9, which equals 100 square yards.

2. (b) If the roll of carpet comes in 12-foot rolls, calculate $30' \times 12$, which equals 360 square feet. Divide square feet by 9, which equals 40 square yards. Double this to take care of the 20-foot width.

Calculate Cubic Yards Answers/Solutions:

1. (b) $L \times W \times D \div 27$. The $L \times W$ or square feet, has been provided. Multiply the square feet of 875 by 0.5 ($6" \div 12 = .5$) and divide by 27. The answer is 16.20.

2. (c) $L \times W \times D \div 27$. Begin by determining the area ($L \times W$) of the tank's base. The formula for computing the area of a circle is pi R squared. The radius is half of the diameter or 9'. Pi (3.14) multiplied by the radius squared (81) is 254.34. Multiply the area of $254.34 \times 32'$ for a total of 8,138.88. Divide 8,138.88 by 27 for a total of 301.44.

3. (c) $L \times W \times D \div 27$. Plug the proper numbers into the formula. The length is 200, the width is 1.17 ($14" \div 12$), and the depth is 1.33 ($16" \div 12$). Multiply $200 \times 1.17 \times 1.33$, and divide by 27. The total cubic yards is 11.52. Round to the next whole number which is 12. Multiply 12 by $65, and the answer is $780.00.

4. (b) $L \times W \times D \div 27$. Plug in the numbers. Length equals 29, and width equals 14. To determine the depth, subtract 96 from 98; the distance between grade and the top of the slab is 2 feet. Subtract the thickness of the slab from 2 feet. $2' - 0.333 = 1.67$. The depth of the sand to be placed is 1.67. Plug this into the formula, and multiply $29 \times 14 \times 1.67$, which equals 678.02. Now, divide this total by 27. The answer is 25.11.

Calculate Cubic Yards Answers/Solutions

1. (d) Begin by sketching an additional 2 feet around the perimeter of the building. Divide the structure into two sections. The first section will measure $54' \times 38'$. The second section will be $24' \times 6'$. Add the square feet of both areas, for a total of 2,196 square feet. This is the $L \times W$ (area) portion of the formula. Next, multiply this by the depth of 3 feet and divide by 27. The cubic yards to be excavated is 244. Take 15% of 244, and add it to the total. The correct answer is 280.6.

2. (b) L × W × D ÷ 27. Begin by multiplying the area (L × W) by the depth of 7 feet: 4,000 × 7 = 28,000. Divide 28,000 by 27 for a total of 1,037.037 cubic yards. Increase this by 20% to account for the swell, which equals 1,244.444. Divide the total of 1,244.444 by 20 (what a truck holds), and the answer is 63 trucks.

3. (c) L × W × D ÷ 27. Plug the proper numbers into the formula. The length is 72, the width is 14, and the depth is 4 feet. Multiply 72 by 14 by 4, and divide by 27. The total cubic yards is 149.33. Divided by 0.90 (100% − 10% = 90%), the answer is 165.925.

4. (a) Begin with adding 5 feet around the perimeter of the building. The new dimensions will be 62' × 43'. Plug the numbers into the formula, L × W × D ÷ 27. 62' × 43' × 2' ÷ 27 = 197.48. Now, divide by 0.85. The answer is 232.33.

Wall Framing Math Answers/Solutions:

1. (c) Quantity = Wall length ÷ on-center spacing + 1. Multiply 25 by 12 to convert the length of the wall to inches. 25 × 12 = 300. Divide the length of the wall by 16. The studs are located every 16 inches. 300 ÷ 16 = 18.75. There are 19 spaces between each stud. Add one stud to the 19 to account for the beginning stud. This wall will require 20 studs.

2. (b) Board feet = Quantity × Name size × L ÷ 12. The quantity is 18, the name size is 2 × 10, the length is 16 feet. 18 × 2 × 10 × 16 = 5760; now divide by 12. 5760 ÷ 12 = 480.

3. (c) Sketch the wall. Draw a dimension line indicating the center of a 3-foot door. The rough opening will be 38½ inches wide. Divide 38.5 by 2 to determine where the jack (trimmer) stud will be placed. 38.5 ÷ 2 = 19.25 inches. Convert the 9 feet to inches. 9 × 12 = 108 inches. Subtract 19.25 from 108 to identify the measurement of where the rough opening will begin. 108 − 19.25 = 88.75 inches.

4. (c) Quantity = LF ÷ on-center spacing + 1. First, convert the 24 feet to inches: 24 × 12 = 288 inches. Next, divide 288 inches by 16 inches. 288 divided by 16 = 18. Add 1 for the first joist: 18 + 1 = 19 joists.

Roofing Math Answers/Solutions:

1. (a) Divide 28 by 2 to determine the middle point of the span. Multiply 14 by 6 because the roof rises 6 inches for each foot: 14 × 6 = 84 inches. The choices provided are in feet; therefore, we must divide 84 by 12 to convert to feet: 84 ÷ 12 = 7. The height of the gable is 7 feet.

Notes

Notes

2. (c) Pitch is a ratio of rise over span. The 4 in ¼ represents the span. We need to determine the slope. A slope is a ratio of rise to run. If 4 is the span, half of 4 will represent the run. ½ is the slope of this building, but we need it in a ratio to units of 12. Place an *x* over 12, and beside it place 1 over 2. Cross multiply: 12 × 1 = 12. Divide 12 by 2. The ratio of rise to run is 6. The answer is 6:12.

3. (d) Run is half of the span: 29 ÷ 2 = 14.5.

4. (b) Begin by determining the gable height: 32 ÷ two will indicate the center point (ridge) to be 16. Multiply 16 by 8 (the roof is rising 8 inches for each foot) to determine that gable height is 128 inches. Convert the 128 inches to feet: 128 ÷ 12 = 10.67. Multiply 16 by 10.67 to determine the area: 10.67 × 16 = 170.72. This accounts for the total area of the gable.

5. (b) The rafters will run to the ridge. Half of the 20 foot width (to the ridge) is 10 feet. Add 1 foot for the overhang: 11 × 1.054 (roof multiplier for a 4:12 slope) = 11.594. Round to the next even number: the answer is 12.

6. (c) Add 1 foot of overhang for each side of the roof: this equals 32 feet. Multiply 32 by the roof multiplier for a 6 in 12 slope roof. This multiplier is 1.118: 32 × 1.118 = 35.776.

7. (b) C = square root of A squared plus B squared. A = 13.5 (half of 27), B = 6: 13.5 × 13.5 = 182.25, 6 × 6 = 36. Add these two numbers: 182.25 + 36 = 218.25. Press the square root symbol, and 14.77 is returned.

8. (d) Begin by sketching the overhang around the perimeter of the building. Section A's new dimensions will be 77' × 32' Calculate the area of this section: $77 \times 32 = 2{,}464$ square feet. To determine the dimension of section B, begin with the 75-foot length of the overall measurement for the building and subtract 39 feet and 20 feet. Section B is 16 feet in width. Section B's new dimensions, with overhang, will be 18 feet × 20 feet. Calculate the area of this section: $18 \times 20 = 360$. Add the two sections together: $2{,}464 + 360 = 2{,}824$ square feet. Increase this by the multiplier, for a 4 in 12 slope. The multiplier is 1.054: $2{,}824 \times 1.054 = 2{,}976.496$. Add 10% for waste: $2{,}976.496 + 297.6496 = 3{,}274.1456$. The closest answer is 3,274 square feet.

9. (d) Begin by dividing 16 by 2 to identify the run for this section. Add the overhang: $8 + 1 = 9$. Multiply by the appropriate multiplier. The multiplier for a 6 in 12 slope is 1.118: $9 \times 1.118 = 10.062$. Although 10 feet might work, you are required to round to the next even number. The answer is 12.

10. (d) Begin by sketching the overhang around the perimeter of the building. Section A's new dimensions will be 77 feet × 32 feet Calculate the area of this section: $77 \times 32 = 2{,}464$ square feet. Section B is 16 feet in width. Section B's new dimensions, with overhang, will be 18 feet × 20 feet. Calculate the area of this section: $18 \times 20 = 360$. Add the two sections together: $2{,}464 + 360 = 2{,}824$ square feet. Increase this by the multiplier, for a 5 in 12 slope. The multiplier is 1.083: $2{,}824 \times 1.083 = 3{,}058.392$. Add 5% for waste: $3{,}058.392 + 152.9196 = 3{,}211.3116$. Divide this total by 100 to convert the square feet to squares: $3{,}211.3116 \div 100 = 32.1131$. If there are three bundles needed per square, multiply the squares by 3: $32.1131 \times 3 = 96.339$. 97 bundles of shingles.

Notes

Notes

Part 2: Administration Answer Key

1.	d	IRC 106.2	IBC 106.2
2.	d	IRC 105.5	IBC 105.5
3.	a	IRC 106.3.1	IBC 106.3.1
4.	b	IRC 105.7	IBC 105.7
5.	c	IRC 105.2	IBC 105.2
6.	d	IRC 111.2	IBC 111.2
7.	d	IRC 110.3	IBC 110.2
8.	c	IRC 114.1	IBC 114.2
9.	b	IRC 106.5	IBC 106.5
10.	c	IRC 106.5	IBC 106.5

Part 3: Design and Planning Answer Key

1. a	IRC 310.1.3	IBC 1025.2.1	
2. c	IRC 309.1	IBC 406.1.4	
3. c	IRC Table 301.5	IBC Table 1607.1	
4. b	IRC 312.2	IBC 1012.3	
5. d	IRC 311.6.1	IBC 1010.2	
6. d	IRC 319	IBC 2304.11.2.1	
7 b	IRC 311.5.2	IBC 1009.2	
8. d	IRC 315.2	IBC 803.1	
9. b	IRC 304.3	IBC 1208.1	
10. d	IRC 303.1	IBC 1205.2 (8% of 100 square feet)	
11. c	IRC 305.1	IBC 1208.2	
12. b	IRC 314.2.6	IBC 2604.2	
13. a	IRC 310.1.1	IBC 1025.5.1	
14. c	IRC 315.1	IBC 803.1	
15. b	IRC 310.2	IBC 1025.2	
16. c	IRC 309.4	IBC 406.1.3	
17. a	IRC 311.3	IBC 1020.1	
18. b	IRC 312.1	IBC 1012.1	
19. b	IRC 311.5.4	IBC 1009.4	
20. b	IRC 309.2	IBC 406.1.4	
21. d	IRC 308.4 #11	IBC 2406.3 #11	
22. c	IRC 311.5.6.1	IBC 1009.11.1	
23. b	IRC 314.2.3	IBC 2603.4.1.6	
24. b	IRC 319.1	IBC 2304.11.2.5	
25. c	IRC 317.2.2	IBC 704.11.1	

Notes

Part 4: Foundations Answer Key

1. c	IRC Table 404.1.1(1)	IBC Table 1805.5 (1)
2. c	IRC Table 401.4.1	IBC Table 1804.2
3. b	IRC 404.1.6	IBC
4. d	IRC Table 402.2	IBC Table 1904.2.2(2)
5. d	IRC 403.1.4	IBC 1805.2
6. a	IRC 403.1.6	IBC 2308.6
7. a	IRC 401.3	IBC 1803.3
8. a	IRC 408.2	IBC 1203.3.1
9. b	IRC 405.1	IBC 1807.4.2
10. c	IRC Table 402.2	IBC 1805.4.2.1
11. b	IRC 403.1.5	IBC 1805.1
12. a	IRC 403.1.4.1 exception	IBC 1805.2.1 exception
13. b	IRC 404.1.5	IBC 1805.5.1.1
14. b	IRC 404.1.5.1	IBC 1805.5.7
15. d	IRC 404.1.8	IBC 1805.5.1.3

Part 5: Floors Answer Key

1. b	IRC 502.6.1	IBC 2308.8.2	
2. b	IRC 502.10	IBC 2308.8.3	
3. d	IRC 502.3.1(2)	IBC 2308.8(2)	
4. b	IRC 502.8.1	IBC 2308.8.2	
5. c	IRC 502.12	IBC 717.3.3	
6. c	IRC 502.10	IBC 2308.8.3	
7. b	IRC 506.1	IBC 1911.1	
8. a	IRC Table 503.2.1(1)	IBC Table 2304.7(3)	
9. d	IRC Table 502.3.1(1)	IBC Table 2308.8(1)	
10. d	IRC 502.6	IBC 2308.7	
11. c	IRC 502.6.2	IBC 2308.8.2	
12. b	IRC 502.8.1	IBC 2308.8.2	
13. a	IRC Table 502.5(2)	IBC Table 2308.9.6	
14. c	IRC 506.2.3	IBC 1911.1	
15. a	IRC 503.3.2	IBC 2303.1.7.1	
16. b	IRC Table 502.3.1(1)	IBC Table 2308.8(1)	
17. d	IRC 502.8.1	IBC 2308.8.2	
18. c	IRC Table 502.3.1(1)	IBC Table 2308.8(1)	
19. b	IRC Table 502.3.1(1)	IBC Table 2308.8(1)	
20. a	IRC 502.8.1	IBC 2308.8.2	
21. b	IRC Table 502.5(2)	IBC Table 2308.9.6	
22. b	IRC 502.7.1	IBC 2308.8.5	
23. d	IRC 502.11.3	IBC 2308.10.7.3	
24. d	IRC 506.2.3	IBC 1911.1	
25. c	IRC 502.3.1	IBC Table 2308.8(1)	

Notes

Part 6: Walls Answer Key

1. b	IRC 602.3.3	IBC 2308.9.2.2
2. c	IRC Table 602.3(5)	IBC Table 2308.9.1
3. c	IRC 603.3.3	IBC 2211.2.2
4. b	IRC Table 602.3(4)	IBC Table 2308.9.3(5)
5. c	IRC 602.8	IBC 717.2.2
6. c	IRC 602.9	IBC 2308.9.4
7. d	IRC 602.6.1	IBC 2308.9.8
8. d	IRC 602.6	IBC 2308.9.11
9. b	IRC Table 603.2(2)	IBC 2211.2.2
10. c	IRC 603.2.4	IBC 2211.2.2
11. b	IRC Table 603.2(1)	IBC 2211.2.2
12. a	IRC 602.11.3 #1	IBC 2308.11.3.2
13. c	IRC 602.10.1.1	IBC 2308.3.1
14. a	IRC 602.6	IBC 2308.9.11 (40% of 3.5")
15. c	IRC Table 602.3(5)	IBC Table 2308.9.1
16. b	IRC Table 602.3(4)	IBC Table 2308.9.3(5)
17. a	IRC 602.7.2	IBC 2308.9.7
18. b	IRC 602.5	IBC 2308.9.2.3
19. a	IRC 602.9	IBC 2308.9.4
20. c	IRC 602.3.2	IBC 2211.2.2.1
21. d	IRC 602.3.2	IBC 2211.2.2.1
22. a	IRC Table 602.3(1)	IBC Table 2304.9.1
23. b	IRC 602.8.1	IBC 717.2.1
24. b	IRC 602.6	IBC 2308.9.10
25. b	IRC 602.3.1	IBC 2308.9.1

Part 7: IRC 2003 Wall Coverings Answer Key

1. b   R702.2.1
2. d   R703.5.3
3. a   R703.4 Table
4. b   R703.6.2.1
5. c   R702.2
6. c   R703.8 #3
7. a   R702.4.2
8. d   R702.3.5 Table
9. d   1926.56 (Table D-3)
10. b   R703.7.3

IBC 2003 Wall Coverings Answer Key

1. d   1405.6
2. c   1405.2 Table
3. d   1405.11.1
4. b   1405.9.1
5. a   2502 Definition
6. c   1405.4 #3
7. b   1405.13.1
8. a   1405.13.1
9. d   1405.3
10. d   1405.17.2

Notes

Notes

Part 8: Roof Framing Answer Key

1. c	IRC 802.8.1	IBC 2308.8.5
2. b	IRC Table 802.4(1)	IBC Table 2308.10.2(1)
3. b	IRC 802.6	IBC 2308.10.4.1
4. b	IRC Table 802.5.1(1)	IBC Table 2308.10.3(1)
5. b	IRC Table 802.5.1(1)	IBC Table 2308.10.3(1)
6. b	IRC 802.7.1	IBC 2308.10.4.2
7. b	IRC Table 802.5.1(9)	IBC Table 2308.10.4.1
8. c	IRC Table 802.5.1(9)a	IBC Table 2308.10.4.1a
9. b	IRC Table 802.11	IBC Table 2308.10.1
10. b	IRC 802.3.2	IBC 2308.8.2
11. d	IRC 802.5.1	IBC 2308.10.5
12. b	IRC 802.3	IBC 2308.10
13. c	IRC 802.9	IBC 2308.10.4.3
14. c	IRC Table 803.1	IBC Table 2304.7(1)
15. b	IRC 802.7.1	IBC 2308.10.4.2
16. c	IRC 802.5.1	IBC 2308.10.5
17. a	IRC Table 802.4(2)	IBC Table 2308.10.2(2)
18. d	IRC Table 802.4(2)	IBC Table 2308.10.2(2)
19. a	IRC 807.1	IBC 1209.2
20. d	IRC 802.1.3.1	IBC 2303.2
21. a	IRC 806.2	IBC 1203.2
22. b	IRC 802.1.3.5	IBC 2303.2.5
23. c	IRC 802.1.3	IBC 2303.2
24. a	IRC 806.3	IBC 1203.2
25. a	IRC Table 802.4(1)	IBC Table 2308.10.2(1)

Part 9: Roof Coverings Answer Key

		IRC	*IBC*
1.	a	R905.2.2	1507.2.2
2.	d	R905.2.5	1507.2.6
3.	b	R905.2.5	1507.2.6
4.	b	R905.2.6	1507.2.7
5.	c	R905.2.7	1507.2.8
6.	c	R905.2.7.2	1507.2.8.2
7.	b	R905.3.3.2	1507.3.3.2
8.	d	R905.2.8.1	1507.2.9.1
9.	c	R905.2.8.2 exception	1507.2.9.2 exception
10.	a	R905.3.2	1507.3.2
11.	c	R905.3.6	1507.3.6
12.	d	R905.3.7	1507.3.8
13.	a	R905.4.3	1507.5.3
14.	c	R905.4.6	1507.5.6
15.	a	R905.7.1	1507.8.1
16.	d	R905.6.5 Table	1507.7.5 Table
17.	d	R905.8.6	1507.9.7
18.	b	R905.8.6 Table	1507.9.7 Table
19.	d	R905.10.4	1507.4.4
20.	c	R905.10.3 Table	1507.4.3 Table
21.	b	R905.5.5.2	1507.6.2
22.	d	R905.9.1	1507.10.1
23.	c	R905.14.3	1507.14.3
24.	b	R907.3 exception	1510.3 exception
25.	a	R905.9.2 Table	1507.10.2 Table

Notes

Part 10: Chimneys and Fireplaces Answers Key

		IRC	*IBC*
1.	b	R1001.1.1	2113.2
2.	d	R1001.3	2113.6
3.	c	R1001.7	2113.10
4.	c	R1001.8.1	2113.11.1
5.	b	R1001.6.1	2113.9.1
6.	b	R1001.6	2113.9
7.	a	R1001.9	2113.12
8.	d	R1001.11(1) Table	2113.16(1) Table
9.	c	R1001.14	2113.18
10.	a	R1001.15	2113.19
11.	c	R1001.12.2 Table	2113.16 Table
12.	d	R1001.16	2113.20
13.	b	R1003.2	2111.2
14.	d	R1003.4.1	2111.4.1
15.	a	R1003.6	2111.6
16.	c	R1003.7	2111.7
17.	d	R1003.8	2111.8
18.	b	R1003.8.1	2111.8.1
19.	a	R1003.9.2 exception	2111.9.2 exception
20.	d	R1003.10	2111.10
21.	b	R1005.2	2111.14.3
22.	b	R1001.10	2113.14
23.	c	R1001.11(2)	2113.16(2)
24.	a	R1001.15	2113.19
25.	c	R1001.2	2113.5

Part 11: Concrete Answer Key

1. d
2. a
3. c
4. b
5. c
6. d
7. c
8. a
9. c
10. a
11. c
12. d
13. b
14. c
15. c
16. c
17. b
18. d
19. b
20. a
21. c
22. c
23. a
24. a
25. b

Notes

Part 12: Masonry Answer Key

1. b
2. b
3. d
4. d
5. c
6. d
7. c
8. a
9. d
10. d
11. b
12. c
13. a
14. c
15. b
16. b
17. a
18. d
19. d
20. a
21. a
22. b
23. c
24. c
25. a

Part 13: Carpentry Answer Key

1. b
2. d
3. d
4. c
5. b
6. c
7. a
8. c
9. a
10. b
11. a
12. d
13. c
14. c
15. a
16. b
17. b
18. b
19. b
20. b
21. a
22. a
23. a
24. d
25. c

Notes

Part 14:
Project Management Practice Exam 1 Answer Key

1. a
2. c
3. c
4. a
5. c
6. d
7. c
8. a
9. c
10. c
11. d
12. a
13. c
14. d
15. b
16. d
17. d
18. b
19. d
20. d
21. a
22. a
23. a
24. a
25. a

Part 14:
Project Management Practice Exam 2 Answer Key

1. b
2. a
3. b
4. c
5. a
6. b
7. c
8. a
9. b
10. a
11. b
12. a
13. b
14. b
15. c
16. c
17. a
18. b
19. b
20. b
21. b
22. d
23. a
24. a
25. a

Notes

Notes

Part 15: OSHA Practice Exam 1 Answer Key

1. c   1926 Subpart P, Appendix B (Table B-1)
2. b   1926.1053 #13
3. b   1926.302(a)(2)
4. c   1926.651(J)(2)
5. b   1926.102
6. b   1926.24
7. c   1926.651(c)(2)
8. d   1926.651(h)(1)
9. a   1926.56(Table D-3)
10. a   1926.50(d)(2)
11. c   1926. 651(C) (2)
12. c   1926.302(b)(4)
13. c   1926 Subpart P, Appendix A
14. a   1926.55(a)
15. a   1926.501(b)(1)
16. d   1926.300(d)(3)
17. d   1926.25(b)
18. d   1926.1053(b)
19. d   1926.651(c)(2)
20. c   1926 Subpart P, Appendix A
21. b   1926 Subpart P, Appendix A
22. b   1926.1053(b)(1)
23. b   1926 Subpart P, Appendix A
24. b   1926 Subpart P, Appendix A
25. d   1926.28

Part 15: OSHA Practice Exam 2 Answer Key

1. b   1926.52 (Table D-2)
2. c   1926.1052(c)(4)iii
3. c   1926.151(c)(5)
4. d   1926.302(e)(2)
5. c   1926.104(f)
6. d   1926.451(e)(2)ii
7. c   1926.250(b)(8)iv
8. c   1926.252(b)
9. b   1926.604(a)(2)i
10. c   1926.451(e)(6)v
11. d   1926.150(c)(1)vi
12. c   1926.104(d)
13. c   1926.250(b)(6)
14. b   1926.50(d)(2)
15. c   1926.104(d)
16. b   1926.251(d)(4)iv
17. c   1926.300(d)(3)
18. c   1926.252(a)
19. d   1926.105(a)
20. c   1926.452(k)(1)
21. c   1926.502(b)(3)
22. c   1926.105(c)(1)
23. a   1926.451(f)(6)(Table)
24. b   1926.1053(b)(1)
25. c   1926.451(e)(3)ii

Part 16: Comprehensive Final Answer Key

1. b	IRC 602.3.1	IBC 2308.9.1
2. a	IRC 403.1.6	IBC 2308.6
3. b		
4. a	IRC 105.7	IBC 105.7
5. a		
6. a		
7. b	R1005.2	2111.14.3
8. d	R905.3.7	1507.3.8
9. b	1926.302(a)(2)	
10. d		
11. b	IRC Table 805.5.1(1)	IBC Table 2308.10.3(1)
12. a	R1003.6	2111.6
13. c	IRC Table 301.5	IBC Table 1607.1
14. c	IRC 309.4	IBC 406.1.3
15. c		
16. c		
17. a	IRC 806.2	IBC 1203.2
18. d	R905.8.6	1507.9.7
19. c		
20. c	1926 Subpart P, Appendix A	
21. c	IRC Table 404.1.1(1)	IBC Table 1805.5(1)
22. b	R1001.6	2113.9
23. d	IRC 105.3.2	IBC 105.3.2
24. b	R905.2.5	1507.2.6
25. d	IRC 315.2	IBC 803.1
26. d	IRC 502.6	IBC 2308.7
27. d	IRC 506.2.3	IBC 1911.1
28. d	IRC 802.5.1	IBC 2308.10.5
29. c		
30. b		
31. d		
32. d	IRC 502.3.1(2)	IBC 2308.8(2)

33. b	IRC Table 602.3(4)	IBC Table 2308.9.3(5)	**Notes**
34. b	IRC Table 602.3(4)	IBC Table 2308.9.3(5)	
35. c			
36. c	R1001.8.1	2113.11.1	
37. d	IRC 502.8.1	IBC 2308.8.2	
38. c	IRC 602.3.2	IBC 2211.9.2.1	
39. b	IRC Table 802.4(1)	IBC Table 2308.10.2(1)	
40. a	R905.9.2 Table	1507.10.2 Table	
41. c	R1001.12.2 Table	2113.16 Table	
42. c			
43. b	1926 Subpart P, Appendix A		
44. a	IRC 310.1.3	IBC 1025.2.1	
45. c	IRC 311.5.6.1	IBC 1009.11.1	
46. b	IRC 405.1	IBC 1807.3	
47. a	IRC Table 502.5(2)	IBC Table 2308.9.6	
48. c	IRC 603.2.4	IBC 2211.2.2	
49. a	IRC Table 802.4(2)	IBC Table 2308.10.2(2)	
50. d	R905.2.5	1507.2.6	
51. a			
52. b	R1001.1.1	2113.2	
53. a	R1001.15	2113.19	
54. a			
55. b			
56. c			
57. d	1926.25		
58. d	IRC 319	IBC 2304.11.2.1	
59. b	IRC 404.1.6	IBC	
60. a	IRC Table 503.2.1(1)	IBC Table 2304.7(3)	
61. c	IRC 602.10.1.1	IBC 2308.3.1	
62. c	IRC Table 805.5.1(1)-a	IBC Table 2308.10.3(1)-a	
63. b	R905.3.3.2	1507.3.3.2	
64. a	R1003.9.2 exception	2111.9.2 exception	

Notes

65. d		
66. d	IRC 303.1	IBC 1203.4 (8% of 100 square feet)
67. b	IRC 404.1.5.1	IBC 1805.5.7
68. a	IRC 502.8.1	IBC 2308.2
69. b	IRC 602.5	IBC 2308.9.2.3
70. a	IRC Table 802.4(1)	IBC Table 2308.10.2(1)
71. b	R905.5.5.2	1507.6.2
72. c	R1001.14	2113.18
73. d		
74. b		
75. c		
76. d	IRC 106.2	IBC 106.2
77. b	IRC 312.1	IBC 1012.1
78. b	IRC 309.2	IBC 406.1.4
79. a	IRC 403.1.4.1 exception	IBC 1805.2.1 exception
80. c	IRC 502.12	IBC 717.3.3
81. b	IRC 502.7.1	IBC 2308.8.5
82. c	IRC Table 602.3(5)	IBC Table 2308.9.1
83. b	IRC 802.7.1	IBC 2308.10.4.2
84. a	R905.3.2	1507.3.2
85. b	1926.651	
86. b	R1003.2	2111.2
87. a		
88. c		
89. d	IRC 311.6.1	IBC 1010.2
90. a		
91. a	IRC 310.1.1	IBC 1025.2
92. b	IRC 502.6.1	IBC 2308.8.2
93. d	IRC 602.6	IBC 2308.9.11
94. b	IRC 802.7.1	IBC 2308.10.4.2
95. a		

96.	b	R905.8.8 Table	1507.9.7 Table
97.	c	R1001.2	2113.5
98.	c		
99.	a		
100.	c	R905.14.3	1507.14.3

Notes

ABOUT THE AUTHOR

American Contractors Exam Services has successfully prepared candidates for state required licensing exams since 1991. A team of full time instructors comprises field experience, training expertise and formal education to formulate an effective and dynamic approach to concisely present pertinent test information in a seminar format. The seminars are conducted throughout the United States utilizing multimedia and interactive technology.

Utilizing effective graphics, pictures, and bulleted text, complex material is presented in an easy-to-understand format. This simplified approach has led to thousands of success stories and happy clients. One is sure to benefit from any material written or presented by American Contractors Exam Services.